GW00468372

ISBN 978-0-331-08262-3
PIBN 10075120

A TEXT BOOK

ON

MARINE MOTORS,

Being a description of most of the Leading Types of various Marine Motors as now manufactured,

BY

CAPTAIN E. DU BOULAY,

LATE ROYAL ARTILLERY.

WITH 90 ILLUSTRATIONS.

LONDON:

THE 'YACHTSMAN,' 143, STRAND, W.C.

1902.

S. Chick & Co.,
PRINTERS,
16, MORTIMER STREET,
LONDON, W.

PREFACE.

THE following pages are a slightly revised reprint of a series of articles on Marine Motors, which appeared in "THE YACHTSMAN" during the winter 1901-2.

As it is, however, a difficult matter to study any subject from periodical instalments of information, and, as the subject of motors for yachts and boats appears to possess a great interest to so many Bristish yachtsmen, the articles are here offered in the more compact form of a book, which, it is hoped, may prove useful to those who, whilst desirous of providing themselves with some form of propulsion, other than sails, may feel debarred from doing so from a want of knowledge as to the principles and peculiarities of the various motors offered to them by the manufacturers of so many countries.

Some of the happiest days of my life have been spent in sailing, but I have always been one of the first to realise what a valuable adjunct to the sail, a compact motor, always ready at a moment's notice, would be. The recent development of the oil engine appears to provide just such a motor for anyone who will take the trouble to understand it.

E. DU BOULAY.

Ryde, I.W.

TABLE OF CONTENTS.

CHAPTER VIII.

CHAPTER IX.

CHAPTER X.

CHAPTER XI.

CHAPTER XII.

CHAPTER XIII.

LIST OF ILLUSTRATIONS.

LIST OF ILLUSTRATIONS—*continued.*

A TEXT BOOK ON MARINE MOTORS.

CHAPTER I.

SUMMARY.

INTRODUCTORY—PETROLEUM AND PETROL, WHENCE OBTAINED—THEIR NATURES AND PECULIARITIES—EFFICIENCIES OF VARIOUS ENGINES.

NO ONE, unless he is blinded by prejudice, can now fail to acknowledge the enormous influence that the "oil engine" is destined to have on our daily, social, and mercantile life in the immediate future, and even a Mr. Wells can hardly dare to prophesy as to the many changes and reformations that will be brought about by its means. It is already revolutionizing land transport, and there can be no doubt that it will have a great effect on marine transport, and, especially so, as regards the many who now take their pleasure afloat, though at present the latter hardly seem to have grasped the capabilities of the oil engine, as adapted to their requirements. The landsman has had a hard battle to fight, because he has been assailed with all the usual weapons of prejudice, *i e.*, first of all—ridicule, then abuse, and finally, persecution, of a kind that, lucky for the English nation, rarely disgraces its Courts of Justice; but he has persevered through it all, and, let us hope, will soon reap his reward—"fair play."

The yachting-man, on the other hand, has had no struggle at all. The Coastguard do not lurk in unsuspected places, ready to time (with watches innocent of second hands) any motor boat passing from buoy to buoy, so that they may swear to impossible records of speed being made, and thus enable heavy fines to be imposed, (*vide* frontispiece). No; the sea spells freedom, and those who use it are free to adopt any form of propulsion afloat that they please. It cannot therefore be doubted that, very shortly, a mild revolution in yachting will take place, as the utility of the motor becomes known, and it is recognised that its reliability and endurance (in which qualities it was at first lacking) are becoming greater each month. It is only natural that those who use the sea should wish to avail themselves of this motive power, so that whilst retaining all the charm of a sail-propelled vessel, they may have something to fall back upon in a calm, either by means of an engine on board, or by means of a launch light enough to be easily hoisted in the davits of even a small yacht. Moreover, a comparatively poor man can now have a self-propelled craft always ready at a minute's notice to take him about on the water, far cheaper to buy than the smallest steam launch, and far cheaper to work because he can work it himself.

A small yacht of 24 or 25 ft. waterline may be fitted with a motor and yet retain all the room available in such a boat without an engine. The motor will stow below the cockpit and will not interfere with good cabin accommodation. The true single-hander will be found in this type, for the single-hander must, unless he wishes to be a slave to his hobby, make a port every night. Then, if called upon suddenly to shift his berth, it is less trouble to start the engine than to set the sails. He need not worry to beat through a tortuous channel, and dodge obstructions with this aid under him. In very much larger yachts the same advantages, of course, will apply, and it is astonishing how very easily a well-shaped craft, even of large displacement, can be moved, provided the power is continuous, and not spasmodic, as in towing with a dinghy.

In classifying the various forms of motors to be described, it may be said that the ordinary steam engine will not be touched upon, as there are so many excellent works already published thereon, and moreover fully qualified engineers are always obtainable for this branch of mechanics, whilst the opposite is the case as regards the other and newer types of motors. The motors to be described therefore will be the following:—(1) Heavy oil engines, (2) light oil engines, (3) steam engines with oil fuel, (4) electric engines.

Now, as three out of the four types as above use 'oil,' and seeing that the battle of public opinion seems to rage about the relative dangers and advantages of engines using heavy or light oil (quite a burning point in fact), it will be necessary to go more or less fully into the nature and properties of these oils, before going into details concerning the engines using them.

Practically all the oils that we use for these engines are derived from Petroleum or Rock Oil, which, like coal, is found ready made for us by nature, in a great many countries of the world; our chief supply comes from America and Russia.

This petroleum consists of a great many different substances all mixed together, and the first thing the manufacturer does is to fractionally distil it, that is to say, by gradually warming it; first of all the light volatile oils pass off as vapour, and are condensed into suitable receptacles. On an average, American Petroleum contains about 16 per cent. of these oils, whilst Russian Petroleum only gives 4 per cent.; as the heat of distillation is gradually increased, the next substance to come away is what the Americans call kerosine and we here paraffin, that is the oil we use in our ordinary household lamps. American Petroleum gives about 50 per cent. of this oil, whilst Russian only gives 27 per cent. As the heat of distillation is further increased, the heavy lubricating oils that we use for the cylinders of gas and oil engines, vaseline, and paraffin wax, &c., gradually come away and are separately stored for use.

The light oils are now again distilled, and ultimately what we use here for engines, and known as Pratt's Motor Spirit or Petrol, is obtained and sent over here in air-tight tins. These contain two gallons and are sold at an average price of 1/4 per gallon.

The two oils that we are concerned with here are petrol, as we shall call it, and paraffin oil. Petrol as we get it here is of an average density of ·680 at a temperature of 60°F. (10 gallons of water weigh 100 lbs., and 10 gallons of petrol 68 lbs.), and it gives off an inflammable vapour at any temperature above freezing point; and as this peculiarity has a great and important bearing on its use in motors we must explain more fully what this means: it means that it is always trying to evaporate into the surrounding air, and must therefore be always kept in closed vesssels, and if some of it is spilt on anything (even if the air is freezing), a lighted match, or a hot flaming electric spark held above it will set fire to the vapour given off by it. This peculiarity is what makes it so useful for motors, because the engine only has to draw into its cylinder a supply of air at the ordinary temperature, over, or through a certain quantity of petrol, and an explosive mixture is then immediately ready there to be fired at the proper moment in order to give out energy or 'do work.' On the other hand this peculiarity renders it always necessary to use it with caution, because a leaky joint, or valve, might allow a certain quantity to escape and perhaps accumulate under the platform or bottom boards of a yacht or boat, and we should then have an explosive mixture all ready to be fired by any naked flame, or extra hot electric spark (an ordinary electric spark would not so fire it). In motor cars it can be arranged so that any overflow of petrol may fall on to the road, but in a yacht or boat this cannot be done, and consequently great care should be taken that no such leak can take place.

That this risk is not a great one is proved by the fact that hundreds of these petrol motors are now being used afloat in various countries, and we rarely hear of accidents, but still there is always a certain element of danger attendant on the use of petrol afloat, which does not exist in the use of paraffin.

Speaking generally, it may be said that if the oil engine is to be in the hands of unskilled persons (such as fishermen, for instance), or if the storage tanks and connecting pipes have to be below decks, as in the case of a sea-going yacht, then it would probably be preferable to use a paraffin engine, but in the case of ordinary open launches and boats, and under the care of reasonably skilled persons, then the petrol engine possesses certain great advantages over the other, and these advantages will render it more popular for launches and boats, in spite of risk.

In a further chapter, a plan will be shown for using a petrol engine, which has been adopted by the author in a 5-ton sailing boat, which may be said to avoid all danger in the use of petrol; and a suggested plan for a 16-ft. open fishing boat will also be shown, with the same immunity from risk.

Now, as regards paraffin; this oil is of an average density of ·82 (*i.e.*, 10 gallons will weigh 82 lbs.) and, it is not allowed to be sold in England with a lower flashing point than 73°F., that is to say, if it gives off an inflammable vapour at any temperature under 73°F. Many of the paraffin oils sold here would have to be warmed up to 80°F. and over before they would give off any inflammable vapour. It will be seen, therefore, that this peculiarity renders its use in oil engines really practically quite safe in the hands of more or less unskilled persons, whilst no one need object to having it stored below decks in a yacht.

On the other hand, it will be seen and understood that, before it can be used in an oil engine, hot air must be mixed with it in order to form the requisite explosive mixture, which mixture must be also kept hot until the actual moment of firing, or it would condense again; and thus a more or less powerful lamp is required with these engines to keep up the supply of heat. Certain paraffin engines are constructed so as to keep up themselves the supply of heat necessary to both vaporise the oil and to fire it, too, when once they have been fairly started by means of the lamp, but in the small sizes it is very difficult to get the necessary reliability; and, moreover, if the engine is stopped from any cause, the lamp has to be started again in order to start the engine.

Owing to the fact of the vapour being highly heated before it is drawn into the cylinder a charge does not contain so much explosive mixture as in the case of a petrol engine, which draws in the charge quite cold, so the same cylinder does not give out as much power with paraffin as it does with petrol. Experiments have shown that, in order to get the same power with a paraffin and a petrol engine, the piston displacement, *i.e.*, the area of the piston multiplied by its stroke, must be in the proportion of 100 to 78. Another point is that, as paraffin does not evaporate into the air as petrol does, it is very liable to creep all about different parts of the engine, just as it does all about our household lamps; and this causes the engine to smell a good deal as it becomes heated, at first, until the oil has been evaporated off it.

With these drawbacks, however, the paraffin oil engine, as made at present, is a good, reliable and useful engine that will always hold its own for certain purposes, whilst it is not at all impossible that certain recent improvements as carried out by men like Diesel and Banki might at any time assume a practical form, which would perhaps render it more popular than its petrol rival; because paraffin has a great advantage over petrol in the way of cheapness, and, moreover, it can now be obtained in almost every village, while petrol, of course, is not so easily procurable. This is a very important point to be considered by those who intend to cruise in out-of-the-way waters.

Both the Diesel and Banki paraffin engines work with a very heavy compression—(about 600 lbs. to the square inch) and this compression ignites the charge without any igniting apparatus. Great economy of fuel is obtained by this heavy compression; in fact, these two motors are the most efficient that we know, giving out as power no less than 28 to 30 % of the calorific value of paraffin oil. Steam engines give out from 1 to 8 % of the calorific value of their fuel, according to their size and system, and the best oil engines, as generally used, give 16 %. With these latter engines we have, as a matter of fact, to use one pint of petrol and rather more one than pint of paraffin to get 1-horse power for an hour.

Having now given this description of the fuels, we will give in the next chapter an account of the various motors in which they are burnt in order to produce power.

CHAPTER II.

SUMMARY.

ENGINES USING PARAFFIN OIL—PRINCIPLES ON WHICH THEY WORK—FOUR AND TWO CYCLE SYSTEMS—IGNITION—LUBRICATION—POINTS TO LOOK AFTER.

HEAVY oil engines are the engines using paraffin oil, as we explained in the previous chapter; and the general construction and principle on which they work will be gathered from the diagram shown in Fig. 1.

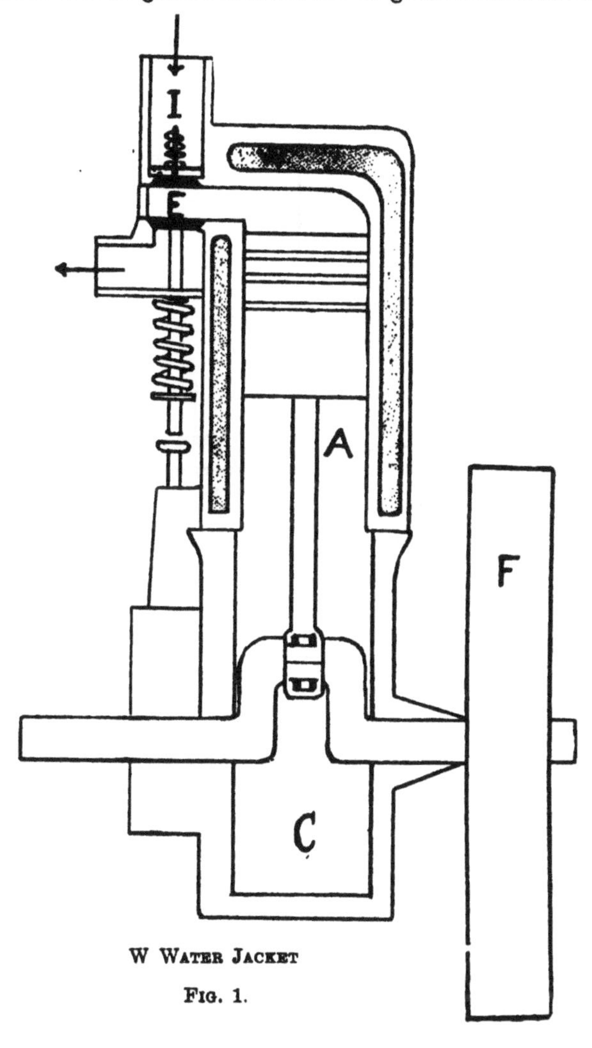

W WATER JACKET

FIG. 1.

A crank chamber *c* of cast iron, or in the later patterns of aluminium for lightness, contains the crank shaft which revolves within long bearings of gun-metal or preferably phosphor bronze; and on the outer end of the crank shaft is keyed the flywheel F. On the top of the crank chamber is bolted the cast iron cylinder A, within which the piston works and which

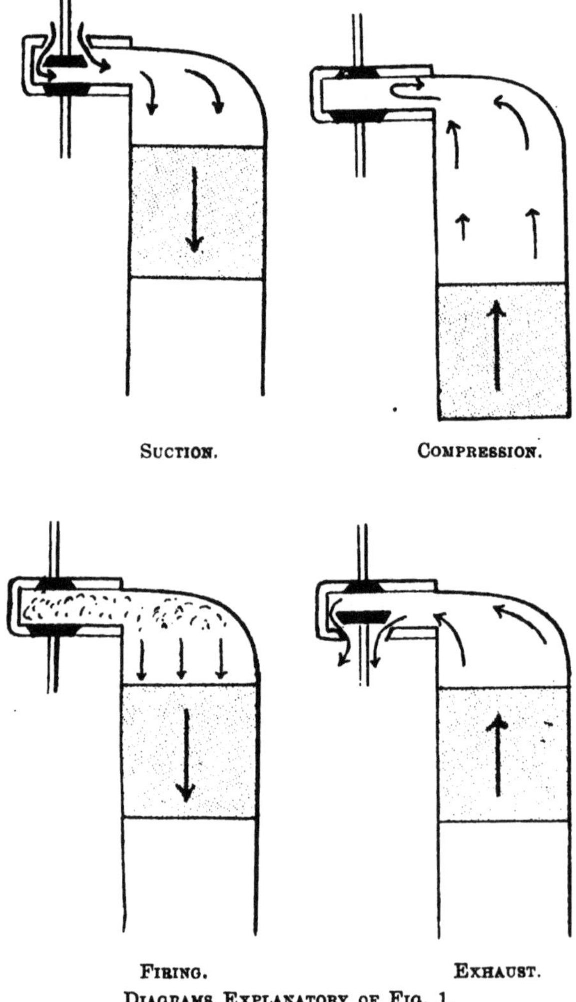

DIAGRAMS EXPLANATORY OF FIG. 1.

piston is connected to the crank by a short connecting rod, the piston is fitted with three or four piston rings, which must be made a good fit in order to stand the pressure and heat. The top of the cylinder is called the combustion chamber, and opening into it is a recess which contains the two valves

I and E: I is the inlet valve and E is the exhaust valve. It will be noticed that these valves are of what is called the mushroom type, and that they both open inwards, they are held pressed against their seats by spiral springs. As the repeated combustion within the cylinder would soon overheat it and the piston it is necessary to cool it, and this is done by a water-jacket all round the cylinder and head; and, as the engine works, it continually pumps cooling water through this water-jacket. These being the chief parts of the engine, we will now proceed to explain the principle on which it works, and it may be stated generally that they all work on what is known as the Otto cycle, or four cycle system, that is to say, that there is an explosion within the cylinder at every fourth stroke, or during one half of every other revolution of the fly-wheel. This explains why we do not get the same power out of the cylinder with oil as we should with high pressure steam, because in the latter we should get four impulees, where we only get one in the former.

Supposing the fly-wheel and crank shaft to be revolving, on the first down stroke of the piston, it creates a vacuum in the cylinder above it, and thereby draws in a charge of heated air from the vaporiser, together with a certain admixture of paraffin vapour, this mixture enters the cylinder by pushing down the inlet valve against its spring. On the next upstroke of the piston this inlet valve closes, and the mixture is compressed until at the top of the stroke, it is compressed to about 30 or 40 lbs. or more on the square inch, in the combustion chamber. At this moment (or in reality a little sooner), the compressed mixture is fired by means of the ignition gear—which we will explain later—and the pressure rises to about three or four times the amount, consequently driving the piston downwards, and doing work. Just before the piston reaches its bottom position, the exhaust valve is raised off its seat by means of the exhaust valve gear, to be described later, and the burnt gases rush out through the exhaust pipe into the atmosphere through an exhaust drum or silencer in order to deaden the noise. This is the cycle of operations on which the engine continously works, *i.e.*, suction, compression, firing exhaust, and when in proper order it will go on working just as long as there is any oil to drip into the vaporiser and make the necessary firing mixture. The four diagrams will show the cycle and the action of the valves more clearly.

It will be noticed that the exhaust valve is only opened during every other upstroke, and in order to effect this, a cam is generally placed on a supplementary shaft which is geared to the crank shaft by spur wheels of two to one gear, that is, the spur wheel on the supplementary shaft has twice as many teeth as that on the crank shaft. At the proper moment, the cam raises a rod which presses up the spindle of the exhaust valve, and holds it off its seat for the required period. As this valve has to be raised at first against the pressure in the cylinder, it entails a good deal of work on the lifting cam, and the whole of the lifting gear has to be made very strong. The inlet valve works itself, being forced open by the incoming gases entering the partial vacuum in the cylinder, and it closes itself by the action of its spring, as soon as the piston commences to rise.

The diagram 3 shows in a graphic way the pressure exerted in the cylinder during a working stroke; the mixture has been compressed by the piston to about 30 lbs. on the square inch. As soon as it is fired the pressure at once rises to about 120 lbs. per square inch, and as the piston begins to

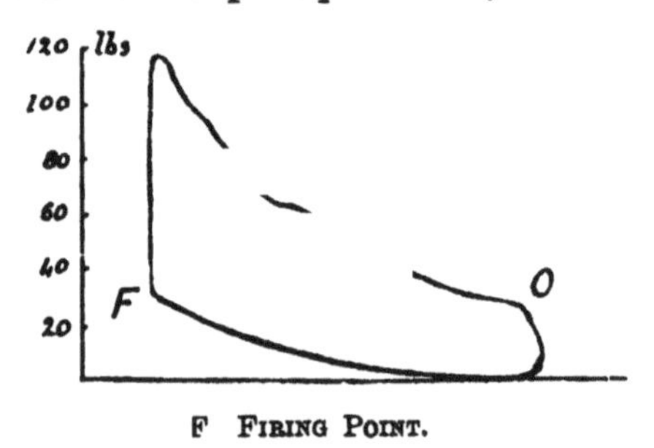

F Firing Point.

O Exhaust Valve Opens.

Fig. 3.

move outwards and thus increase the room in the cylinder, the pressure gradually falls to about 25 lbs., at the point O the exhaust valve is opened by the cam, and the pressure rapidly falls as the burnt gases rush out into the air.

If we calculate out the area of useful work, as shown by this diagram, and multiply it by the number of firing strokes per minute, we get at the number of foot lbs. of work done in the cylinder per minute, and consequently the *indicated horse power*. As, however, a good deal of this power is absorbed in the cycle, especially by friction, &c., *the brake horse power* is always a good deal less; but as this brake horse power, or power really available for turning a screw shaft, or capstan, is all that we get out of the engine, it should always be remembered that in comparing engines or seeing what we are buying, we should only talk about actual brake power, and not about indicated horse power.

In speaking of the moment of ignition, it was stated that it generally took place just before the piston reached its top stroke, and the reason for this is that the ignition of the mixture is not instantaneous, but a certain fraction of time is required for the maximum pressure to be reached; the greatest power is obtained when the maximum pressure is reached just when the piston is at its top stroke, so that to get this, we must commence to ignite it just before the top stroke. In each engine there is a certain point near the end of the compression stroke when the ignition should take place, in order to get the best results, and this point depends on the make of the engine as well as its speed, so that it has to be found out by experiment. Besides this ignition point, the actual power given out by an engine depends a good deal on the actual moment when the exhaust valve

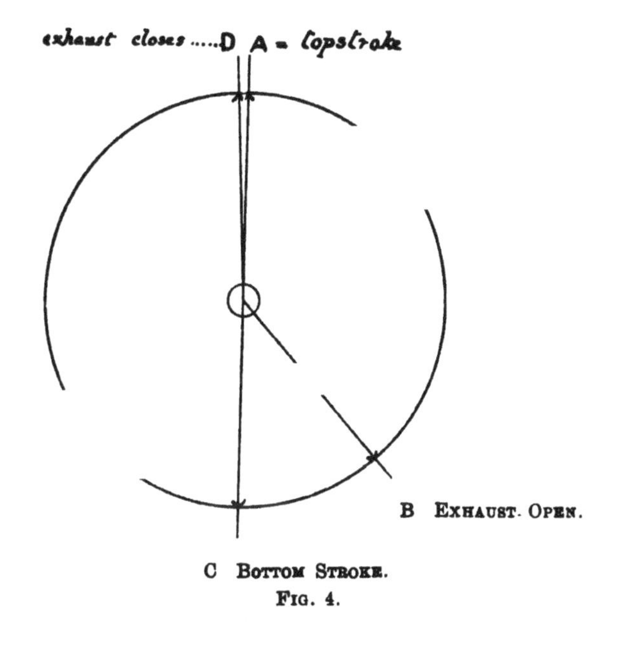

Fig. 4.

begins to open, and when it closes; on the strength of the spring regulating the inlet valve, and also on the amount of lift of both these valves. It will be seen, therefore, that given the same cylinder and stroke, one maker will get more power out of it than will another. Even if an engine is properly adjusted as to these points, when it is first sent out, yet, after a certain time owing to wear, it may want re-adjusting, and it may not be amiss to give a few general rules as to this matter for the use of those owning engines. The valves should only lift about $\frac{1}{4}$ of the diameter of their seating, and the moment of opening and closing of the exhaust valve may be graphically shown, as in Fig. 4, where A is a mark on the fly-wheel, which the owner must make exactly in line with the crank, so that when this mark is at the top or bottom the piston is in a similar position inside the cylinder.

The pump for circulating the water is a very important portion of the engine, as, if it stops working, the cylinder will get very hot, in fact so hot that all the lubricating oil will be burnt up, and the friction will be so great that it may stop the engine altogether, and in such a case it is very likely that the piston rings will suffer. Different makers use different kinds of pumps, but probably the best and simplest form is the rotary pump, which is found to work best with motor cars, as there is very little packing to look after; and if it is driven by gear on the shaft there can be no trouble as from a belt slipping.

For convenience of manufacture the head of the cylinder is nearly always made of a separate casting, which is bolted on to the cylinder, and in selecting an engine preference should always be given to one in which the circulating water cannot find its way into the cylinder through the joint between the cylinder and its head. It is a curious fact that if water gets in thus, it will interfere with the ignition of the charge. Some of the earlier engines had only a thin strip of asbestos paper in this joint between the inside of the cylinder, and any leak through this asbestos packing was sure to interfere with the ignition sooner or later.

We have put off the description of the ignition gear till the last, for it is the most important point of the whole motor, because, in order to get the proper power out of an engine, the ignition must take place at a definite time of each stroke, as was explained just now, whilst if the ignition fails the whole engine is useless. There are two forms of ignition now in general use, *i.e.*, hot tube and electric. As nearly all paraffin engines use the former, we shall defer describing the electric ignition until we come to the petrol engine, and will only now describe the hot tube. It will be remembered that it is necessary that highly-heated air should be used to vaporise the paraffin oil before admission to the cylinder, and as a heating lamp is required for this purpose it is also used to heat the tube. The tube is made of porcelain or of some refractory alloy, or of nickel or platinum, so as to stand being kept up to a bright red heat continuously during the working of the engine, and it must be strong enough to stand the force of explosion inside it when in this heated state. It is screwed into the combustion chamber at a suitable place, so that on the compression stroke the mixture is forced into it and compressed, the outer end of the tube being closed. According to its length, therefore, at a certain part of the compression stroke, the mixture inside the heated tube will reach the bright red part, and it will ignite and force out a flame into the rest of the charge in the combustion chamber, and drive the piston down in its working stroke. The actual length of the tube has to be found out of each engine by experiment, and when thus found the hot tube forms a very reliable form of ignition gear; although not always getting the best power out of an engine, because we cannot vary much the moment of ignition with a tube, though we can with electric ignition. The longer the tube, and the closer the bright red part is to the engine, the sooner the ignition will take place on the compression stroke, whilst the opposite conditions will retard it.

It was stated previously that the best results were got out of an engine by firing the mixture some time before the top stroke of the piston (in some fast running engines with electric ignition the firing actually takes place at half compression stroke), but in starting the engine we have to turn it round by a handle on the fly-wheel, and thus compress the mixture. Now if the tube were arranged to fire much before the top stroke, the effect of the first explosion would be to reverse the motion of the fly-wheel by what is called a back stroke, and this is not always pleasant for the operator. In starting the engine, therefore, the tube must not be too hot, nor must the heated part be too close to the combustion chamber. After the engine is started, however, it will run faster if we can advance the time of ignition by either of the methods mentioned above.

The heat is generally supplied by means of a burner using paraffin oil and the Swedish pattern or Aetna is found to give the best results. Formerly the oil tank for the lamp and the burner were practically fixtures on and part of the engine, but now that such excellent heating lamps of the Swedish pattern can be easily procured, it is far better to have two of these lamps complete all ready to be fixed into their place by a single screw; then, when one from any cause is not burning well, the spare one is held over the burning one, so as to warm it up, and when well alight it can quickly be substituted for the other one without stopping the engine. The principle on which they work is by pumping air into the reservoir containing the oil, so as to force it under pressure into the burner, where the oil is vaporised and burns with a clear violet blue flame without any smoke.

The hot tube stands in the centre of a kind of furnace, with an asbestos lining, and is easily kept up at a bright red heat continuously, whilst an outer iron jacket is arranged so as to prevent the lamp from being blown out, or having spray over it. The walls of the furnace generally constitute the vaporiser for the oil to be used in the cylinder. They are made hollow, and being highly heated they give the necessary supply of heated air, which first of all sprays the oil; and, by subsequently going through other heated chambers, turns it into the proper vapour for complete combustion in the cylinder. The actual apparatus for measuring out the quantity of oil for each charge varies according to the make of engine, and will be described more fully as we come to each type; but when they are properly adjusted, the combustion in the cylinder is really complete; and when once the engine is well heated a white pocket handkerchief can be held in the exhaust for several minutes without showing any marks. The engine will run for months without being opened for cleaning also; but all this depends on the oil feed and vaporiser being accurately designed and adjusted. At first starting, before the engine is well heated, some of the charge will be condensed on the cylinder walls, etc., before it is fired, and this causes a certain amount of smell in the exhaust, besides visible vapour; but these gradually disappear.

With all internal-combustion engines, the lubrication of the cylinder is a vital point as regards their successful action. Owing to the actual presence of flame in the cylinder during its working stroke, the only oils that can be used for lubricating are those derived from petroleum, and known as gas engine cylinder oils, if other oils were thus used they would char into a mass of carbon which would not lubricate the cylinder at all, but would do actual harm to the piston rings. The lubrication is generally effected by keeping a certain amount of oil loose in the crank chamber, so that, at every stroke, the crank dips into it and splashes it all about the crank bearing, the main bearings of the crank shaft, the gudgeon pin at the top of the connecting rod, and on the walls of the cylinder, whence it becomes distributed on the piston rings; as this oil is used up it must be replaced, and this is generally effected by means of a filling hole into the crank chamber closed by a screw plug. The other ordinary working parts of the engine may be lubricated with the usual oil, but it is generally better to only have one sort of oil about, *i.e.*, the cylinder oil, so that there may be no chance of the ordinary oil being introduced into the cylinder, and doing harm.

As the main bearings of the crank shaft become worn, a good deal of the lubricating oil gets forced out, and thus makes a mess, besides wasting the oil. It is, therefore, very important in choosing an engine to see that these main bearings are of a good suitable length and of good metal. Phosphor bronze gives about the best results for these bearings Paraffin engines do not require quite so much cylinder lubrication as do petrol engines, because the paraffin oil acts somewhat as a lubricant itself which petrol does not.

Periodically, the inlet and exhaust valves will require looking at, and if they do not keep airtight they will have to be ground in by turning them round and round with a screw-driver inserted into the slot in the head of the valve made for the purpose, having first put under the seating of the valve a little oilstone dust and water. When the valve is well ground all round its seating, all the oilstone dust should be carefully removed, and the faces of the metals burnished by turning them round and round again with a little paraffin as a lubricant. If thus carefully burnished they will last much longer before requiring to be re-ground. The inlet valve is kept more or less cool by the incoming charges and will not require to be ground as often as the exhaust valve, which, being surrounded as it is by a mass of flame at each exhaust stroke, becomes much more heated and is consequently much more liable to be cut and abraded.

The spring of the exhaust valve must be a good deal stronger than that of the inlet valve, so that, on the suction stroke, it shall never open and admit any of the exhaust gases into the cylinder. Owing to its getting so much hotter, it is more likely to get broken than is the other one, but in a properly designed engine these springs are always easily removable, and can be soon replaced by the spare ones which should always be carried.

This circulation of the water should be so arranged that the water coming away from the top of the cylinder is about blood heat; but in this respect, as in so many others, particular engines seem to have idiosyncracies, which must be studied in order to get the best results out of them. So that all the foregoing hints should be taken more as such, than as accurate rules to be strictly observed in every case.

Some engines are made with the crank chamber quite open, as in the case of an ordinary marine steam engine. But it is better to have the closed crank chamber, both on account of getting better lubrication for the cylinder and piston rings, and also because any gas that leaks past the piston is confined in the closed crank chamber, and does not consequently escape into the air and cause any smell in the vessel. Engines with closed crank chambers are also rather less noisy than those with open cranks; and the only drawback that the closed crank chamber possesses is that it renders the accurate adjustment and tightening up of the crank brasses rather difficult, owing to the confined space.

With these few remarks as to the general principles on which paraffin engines work, we shall, in the next chapter or two, more fully describe certain of such engines representing the chief different types now offered to the public.

CHAPTER III.

SUMMARY.

VARIOUS MAKES OF PARAFFIN FOUR-CYCLE ENGINES—VOSPER—CAPITAINE—HILLIER—SEAL AND MITCHAM ENGINES, AND THEIR PECULIAR FEATURES—PARAGON TWO-CYCLE ENGINES.

We will now proceed to describe some of the paraffin engines that are at the present time before the public here, and will begin with that of Messrs. Vosper, of Portsmouth, as this engine was about the first to be placed on the market.

Fig. 5.
Vosper Engine.

The single-cylinder type is shown in Fig. 5, and it will be seen that it is very compact in its appearance. The single-cylinder type is made for sizes from ½ b.h.p. to 6 b.h.p., and beyond that size up to 20 b.h.p.; it is made with two or four cylinders. The ½ h.p. for dinghies, weighs 1½ cwt., and the 6 h.p., 12 cwt. It has the ordinary vertical cylinder, with water jacket and closed crank chamber, and the two to one gear that opens the exhaust valve also works a circulating pump for the cooling water, and a small pump for raising the oil up into the small glass cup, A, and at the same time causes the feed rod to work backwards and forwards in its guide. The oil pump is designed to pump up more oil than the engine uses, and the overflow runs back into the tank, which may be placed right away from the engine. The principle of the oil feed is that shown in detail in Fig. 6, and known as Roots' feed. it will be seen that the feed rod has a small groove filed out in it, and as this rod oscillates, this groove comes alternately under the glass

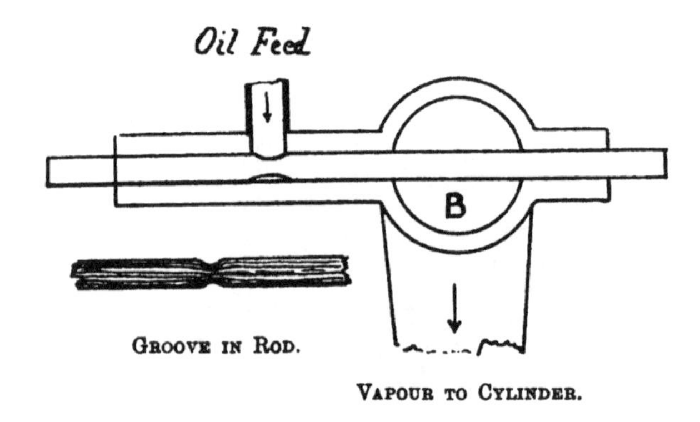

Fig. 6.

oil cup (where it is filled with oil), and then into the air pipe B where the rush of heated air from the vaporiser going into the cylinder sweeps off the oil and vaporises it. The mixture goes through another part of the heated vaporiser, and becomes more perfectly vaporised and mixed with the air, so as to produce complete combustion in the cylinder. It will be noted that the oil feed is a fixed quantity and cannot be varied, and as these paraffin engines usually require a little more oil at first starting when cold, than they do when warmed up, the difficulty is got over as follows:—As soon as the engine is warmed up, an auxiliary air cock can be opened which admits a small amount of heated air to the mixed charge just before it enters the cylinder; the exact amount of air thus given can be found by experiment to be just what the engine requires, and the best results are thus obtained. The groove in the feed rod requires very accurate adjustment, and as the rod after some time may tend to wear a little loose in its bearing, the groove may require to be slightly reduced; this is done by gutting a little hard solder into the groove, and then filing it out till the best results are obtained.

It was mentioned in a previous article that if the main bearings of the crank shaft wore much, there would be a tendency for the oil in the crank chamber to splash out, and this difficulty is got over in this engine, by the plan shown in Fig. 7. A semi-circular groove is turned in the inside of the metal bush forming the bearing, and a piece of Tuck's packing, of

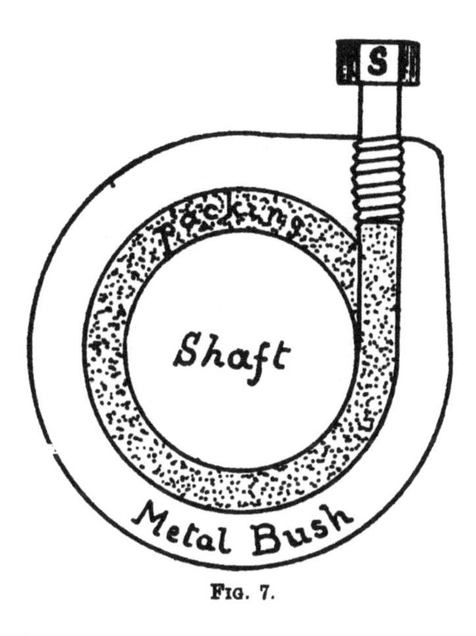

FIG. 7.

suitable size, is forced down by means of the stud S right round the shaft, with which it makes an oil-tight joint, and which can be adjusted from time to time by means of the screwed stud.

The drum with the small pressure gauge on top contains the paraffin oil for the heating lamp, and air is pumped in on the top by means of the small pump to about 15 lbs. per square inch, in order to force it to the burner, where it burns with a clear blue flame, heating the firing tube and the walls of the vaporiser.

An eye is often fitted on the top of a cylinder for slinging the engine in or out of a boat. Fig. 8 shows a larger engine with two cylinders.

A small cock is fitted so as to be able to run away the water from the jacket when wanted, a precaution that should never be neglected when using a boat in the winter.

Fig. 8.

Vosper Engine.

When properly adjusted these engines run well, and are very reliable, and the author has been for hundreds of miles with them both in winter and summer.

Another early type of paraffin oil engine is that shown in Fig. 9, and was brought out by Messrs. Tolch, of Fulham, some years back, under the name of the "Capitaine." As they possess some features quite distinct from all others, it may be worth while describing them. The great feature of these engines is the balanced action and freedom from vibration, owing to there being two pistons and systems of connecting rods, which move in and out together. The outline drawing Fig. 10, shows this balanced action,

Fig. 9.

Tolch Engine.

which works well in practice, so much so that the author, who possesses one, has run it for hours, simply standing on the floor of a workshop and not fastened down in any way. The system of oil feed, however, was rather complicated, and the extra weight of these engines prevented them from becoming as well known and used as their ingenious construction deserved. The same makers also manufacture larger engines, with two vertical cylinders of the usual pattern.

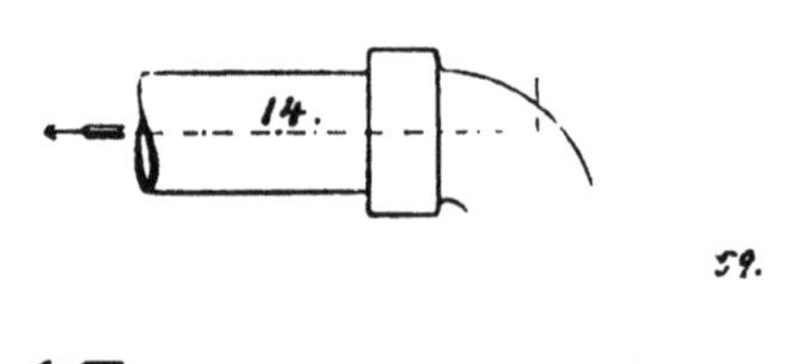

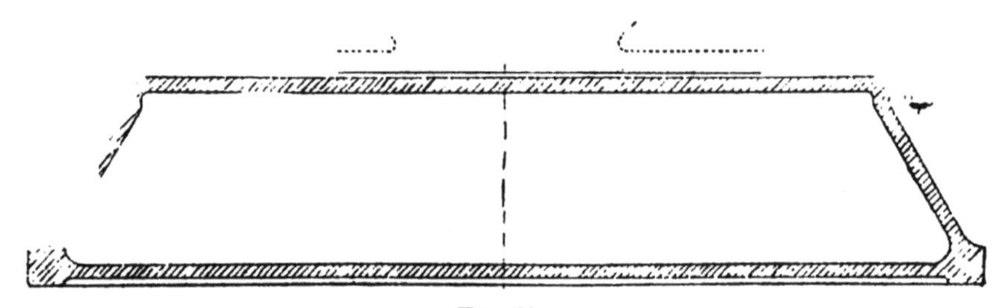

FIG. 10.
TOLCH ENGINE.

Another paraffin engine well worthy of notice is that brought out by Messrs. Hillier, of Romsey, under the name of the "Little Giant," and which they claim to be the lightest in the market for a given power. It is shown in Fig. 11. It certainly possesses great claims on the score of lightness, for a $1\frac{1}{2}$ b.h.p. engine only weighs 150 lbs. They are made in three sizes, *i.e.*, a $\frac{3}{4}$ h.p. with single cylinder, and engines of $1\frac{1}{2}$ and 3 h.p., these two latter having two cyclinders.

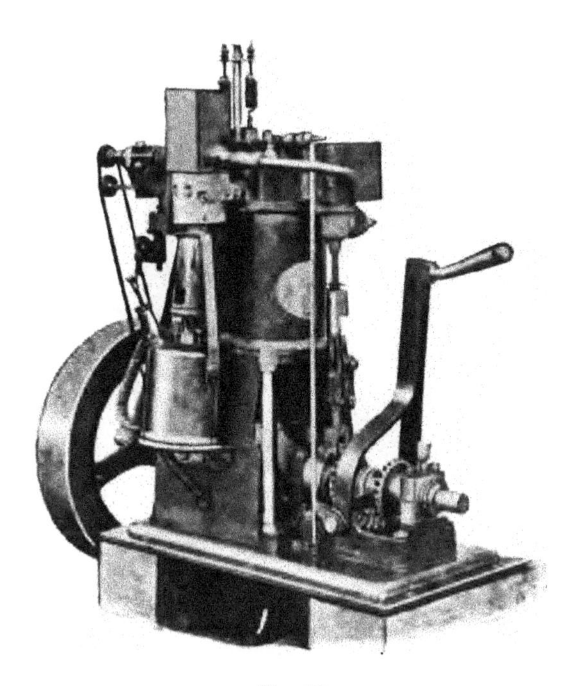

Fig. 11.
Hillier Engine.

They are fired by a hot tube, heated by the lamp as described in a previous article, and shown in Fig. 11; but when once the engine is fairly under way and well heated, the lamp may be put out, and the engine will continue to run by means of its stored up heat, which is maintained by the successive explosions. The oil feed is very ingeniously arranged so that the engine draws up from the supply tank the exact amount it requires for each stroke, and this oil is drawn through the heated vaporiser and mixed with the necessary amount of air in the usual manner before it enters the cylinder. It will be seen, therefore, that no oil pump is required, but only the water circulating pump. The principle of drawing up the supply of oil renders the engine very safe, as there is no oil under pressure (except in the heating lamp at starting), so that if a pipe were broken by accident the oil would not flood the boat. The crank chamber is open, and the propeller is reversed in direction for going astern, and for a temporary stop the engine is thrown out of gear by the same lever as reverses, and shown in Fig. 11. When thrown out of gear the engine is prevented from racing by a governor.

Fig. 12.

Seal Engine.

For sea-going work an iron hood goes right over the engine, to protect it from wind and spray.

These engines run very well, and where lightness is of great importance they are very hard to beat.

Mr. J. Seal, of Strand-on-Green, Chiswick, has recently brought out a new form of paraffin engine for boats, which must be awarded the palm for lightness and compactness, combined with a low centre of gravity, as will be seen by reference to Fig. 12. Its chief peculiarity is that the cylinder is placed upside down, so to speak, *i.e.*, with its closed end downwards, so that the crank shaft is above. The reason for doing this is that as the combustion chamber and lower portion of the cylinder are thus brought so low down in the boat, they are below the waterline, and consequently no pump is required for the circulation of the cooling water, which circulates by being warmed, as in the case of hot water heating apparatus. As soon as the boat moves its motion is made to assist this circulation by the way that the inlet and outlet pipes are arranged. The oil tank is

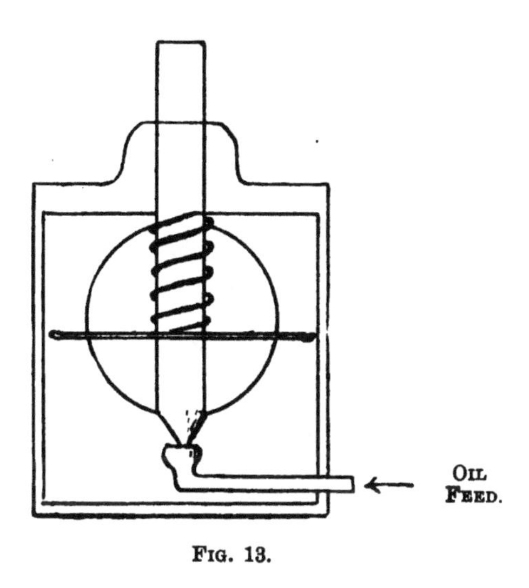

FIG. 13.

carried above the engine (at the other end of the boat), and is led by gravity to the spraying valve shown in Fig. 13. This valve is held down by its weight, assisted by a spring, and the lower end of its spindle forms a small cone, which closes the end of the feed pipe. At each suction stroke of the engine the incoming air raises this valve of its seat and a few drops of paraffin are allowed to escape, and are swept into the cylinder with the air, in the form of spray, through the heated chambers of the vaporiser, which is heated by a heating lamp of the Swedish pattern, and which also heats the firing tube. The exact amount of oil entering at each stroke is regulated by a graduated screw valve fitted between the oil tank and the spraying valve.

The gear for opening the exhaust valve is very ingenious, and is shown in Fig. 14. The rod which opens the exhaust valve carries at its end a revolving wheel cut like a Maltese cross, with alternate deep and shallow teeth. This wheel engages with a spiral cam fitted on the crank axle. As

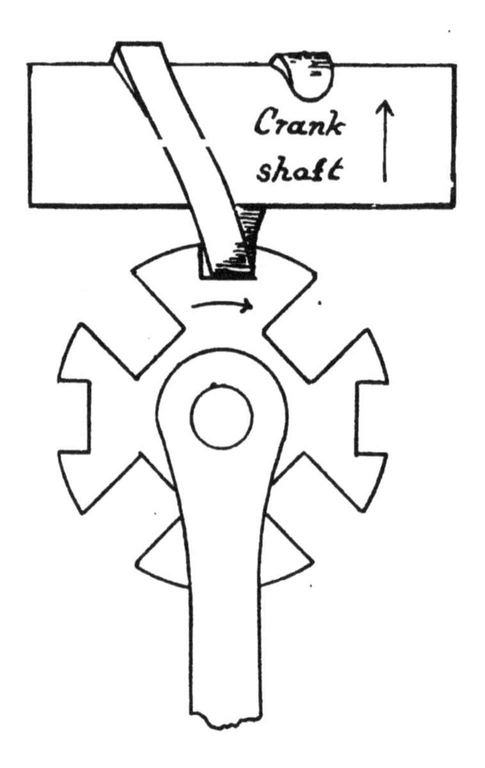

ROD WORKING EXHAUST VALVE.

FIG. 14.

this cam enters a deep tooth on the Maltese cross, it does not push the rod, consequently the exhaust valve is not opened, but when it enters one of the shallow teeth, the rod is pushed away, and opens the exhaust valve. The great thing gained by this is that as the cam which opens the exhaust valve is on the crank shaft (instead of being on another shaft revolving at half the speed, as in other engines) its angular velocity is double that of other engines, and consequently the exhaust valve is opened and closed very quickly, and this assists the engines as far as giving out power is concerned.

A patent igniter is fitted to these engines so that they will continue to vaporise and fire without the heating lamp when once they are well warmed up.

A governor is fitted to prevent the exhaust valve being moved when the engine tries to race, such as when it is disconnected from the screw.

A special form of propeller with reversing blades is also supplied with these engines, which will, with others, be fully described in a future chapter.

These engines are made at present of two sizes, *i.e.*, 1½ and 3 h.p. on the brake, and their weights are respectively only 86 lbs. and 170 lbs., and they stand in the boat 14 and 18 in. high. Owing to their lightness and simplicity, and having no pumps, they can be made as cheaply, if not cheaper, than any other paraffin engine, and are particularly suitable for dinghies of 12 to 14 ft. length.

The Mitcham Motor Co., of Mitcham, Surrey, also supply engines to work with paraffin, but as the same type works with petrol, it will be explained in the article dealing with that type of motor.

There is only one well-known form of oil-engine using paraffin oil, and working of the two-cycle system, *i.e.*, having an explosion at every revolution. It is called the "Paragon," and gives excellent and reliable results when used in a boat.

It is made by Messrs. G. Davies & Co., at Abergavenny, and is offered in three sizes, *i.e.*, 2, 3, and 4, b.h.p. It is naturally somewhat higher and heavier than the modern petrol motors, but nevertheless some owners might prefer it for use afloat.

The ignition is by hot tube, heated by the usual paraffin blow lamp, and the crank chamber is used to compress air, which, as it is driven into the cylinder, forces in with it the proper amount of oil for each charge.

The two-cycle system will be fully gone into and explained later on, when dealing with Petrol engines working on this system, so we will not say any more about it here.

Whilst it is not pretended for an instant that the foregoing list of engines iucludes all those using paraffin oil that are on the market at present, still it is believed that all the chief *types* are represented, and their peculiarities and divergences explained and pointed out, so that anyone on seeing an oil engine, may be able at once to know on what general principle it works if he has carefully studied what has been written on this subject; and, moreover, should he have decided to have a paraffin engine, in preference to a petrol one, it is hoped that he may now be able to form an opinion as to what type would suit his requirements best. Should he not yet have decided the question of paraffin *versus* petrol, it would be better to wait until the different types of petrol engines have been gone into and described, and this question will be begun in the next chapter.

CHAPTER IV.

SUMMARY.

PETROL ENGINES—FOUR AND TWO-CYCLE SYSTEMS EXPLAINED AND COMPARED—HOT TUBE IGNITION—DAIMLER—SIMMS—AND MITCHAM PETROL FOUR-CYCLE ENGINES, AND THEIR PECULIAR FEATURES.

We now have to consider the various types of engines using petrol for fuel instead of paraffin, and, before commencing to describe any of them in detail it must be noted that they are divided into two classes themselves, *i.e.*, those working on the four-cycle system and those working on the two-cycle system. Opinions are very much divided as to the relative value of these two systems; in fact, this question seems to be as burning a one as that of the relative values and advantages of paraffin *versus* petrol. All that we can do here is to describe both systems, pointing out their advantages and disadvantages and leave our readers to form their own opinion as to which type will suit them best.

The four-cycle petrol engine works on exactly the same system (as far as the engine itself is concerned) as the paraffin four-cycle engine, but the means of making the explosive mixture, and firing it, are different.

The two-cycle engine, on the contrary, has an explosion and working stroke at every revolution of the flywheel, and one would naturally expect, at first sight, that it would give out double the power with the same cylinder. The contrary, however, is the case, so that with a given weight and size of engine we get less power with a two-cycle than we do with the four-cycle. This is owing to the explosions not only being weaker, but (owing to the principle on which the old burnt charge is got rid of, and the new one introduced) they are also less frequent, that is, the engine cannot run at the same speed as a four-cycle one. This will be fully explained later on. The two-cycle engine also uses rather more fuel for a given power. There are, however, certain advantages in the two-cycle system which render it particularly useful for marine work, and these may be said to be :—

1. Greater simplicity and cheapness of engine, owing to there being no inlet or exhaust valves to be worked, and periodically ground in.
2. Less vibration and noise, owing to the explosions being weaker, also less noise from the exhaust.
3. Owing to the same cause, the flywheel can be made lighter, and of less diameter, and this latter point has a direct influence on the position of the engine in the boat, as with a smaller flywheel, the engine can be placed lower down in the boat.
4. Greater ease in starting.

Now we must admit that the Americans are a practical nation, especially as regards machinery, and, as a fact, seeing that nearly all their light marine engines are of the two-cycle type, we may say that the advantages, as set forth above, are of real practical value, and that they more than overbalance the loss of power obtainable with a given weight and size and the extra fuel required.

Speaking generally, we may therefore say that where we are putting a moderate power into a boat, such as a small launch, or auxiliary, the two-cycle system would appear to be the best; but if we wish to have a larger launch, or auxiliary which requires more or less full powered engines, then the four-cycle system would probably give better results.

It is rather a curious fact that, until lately, these two-cycle engines were hardly ever seen over here, and the reason most probably is that the only petrol motors that we had seen were those in use in motor cars. Now the two-cycle engine is not in use for this purpose, chiefly because of the loss of power entailed thereby, and the object of motor car makers is to obtain the greatest possible power with a given weight, and they do not care how complicated the engine is, provided they can get the power and probably beat some record, which fact will send the public rushing in to buy that particular car. The earliest petrol engines, therefore, used over here for marine work were the Daimler engines, working on the four-cycle system, and which were the pioneers of petrol. Gradually our makers learnt to copy these engines, and they naturally stuck to the four-cycle system. Certain firms, however, quite realised the advantages that would be obtained by the use of the two-cycle system, and the author has watched with great interest the endeavours of three of them to produce a satisfactory engine on this system; but after spending hundreds of pounds on experiments, all the three firms gave up the idea, and two of them went in for the four-cycle system again. There is something very tricky in the manufacture of two-cycle engines that are to work with reliability, and the least deviation from the proper pattern will spoil this result.

Our cousins across the Atlantic seem to have got over this difficulty, for many of their two-cycle engines are all that can be desired for certain classes of marine work, and they are rapidly coming into use here.

As the four-cycle petrol engine was the first to be used here for marine work, we will commence by describing this pattern before the two-cycle one.

The crank chamber, cylinder, piston and rings, inlet and exhaust valves, and the gear for working the latter, are practically the same as in the paraffin engine, and may be understood by a reference to Fig. 1, chapter 2, but here the similarity ceases, because no heating lamp to the vaporiser is required, but the engine draws in at each suction stroke a charge of air through a carburetter, and, as was explained when treating of petrol, this forms the necessary explosive mixture at once, all ready to be fired by the electric spark, which is generally used for petrol engines.

All the early Daimler engines, however, used the hot tube ignition, the tube being a light nipple of platinum, which was very quickly heated by lamp burning petrol. At first the petrol was fed to the heating lamp by gravity, whilst later the air pressure feed was employed. Figs. 15 and 16 give an illustration of the platinum tube fitted to the plug all ready to be screwed into the combustion chamber, and the heating lamps complete for a double cylinder engine.

It will be seen that an ordinary tyre pump forces in air under pressure through the stop valve R, to the drum L containing petrol; the actual pressure is shown on the gauge E, and the petrol is thus forced through the stop cock C through the regulating valves F F to the burners G G, which act on the Bunsen principal, and give two clear blue flames free from soot, keeping the platinum tube at a bright red heat.

As the presence of these flames in a boat was found to be undesirable, this system gradually gave way to that of the electric ignition. This question of electric ignition is a very elaborate one, and will require a chapter all to itself, but we may say that the mixture is fired by a spark being produced at an exact moment of time between two conductors in the combustion chamber. It has been aptly said that the modern oil engine is an electric spark with an engine built round it, and certainly the electric ignition is by far the most important and delicate part of the engine. Some of the great advantages that this system of ignition possesses over the hot tube, may be stated as (1) the relative safety, as there is nothing in it to ignite any petrol that might by accident be spilt about, and (2) the possibility of exactly timing the moment of ignition at any part of the compression stroke. (3) A greater range of ignitable mixtures. This latter advantage renders the petrol motor easier to start than the paraffin

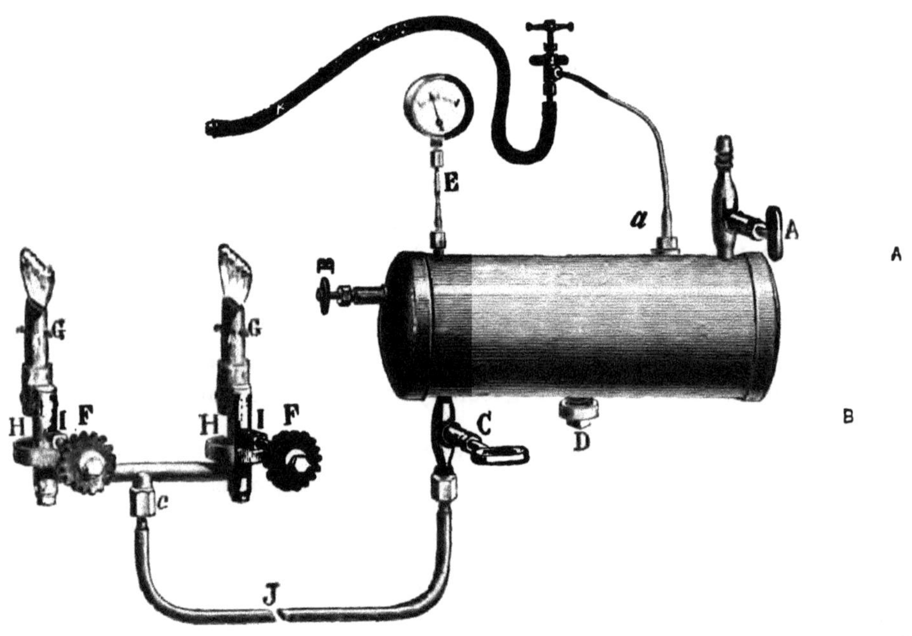

A

B

C

FIG. 15.
RESERVOIR AND BURNERS FOR HOT TUBE IGNITION.

FIG. 16.
PLATINUM IGNITION TUBE.

one, because the hot tube has a much smaller range of ignitable mixtures, and it also renders the ignition more reliable when started, because it is not so liable to misfire should the mixture vary slightly from the theoretically correct one.

The best explosive results are obtained from a mixture of one part of petrol vapour to about 8·3 volumes of air. If the mixture be enriched so that there are four parts of air only to one of petrol vapour, then it will just ignite and burn slowly; but if the mixture be impoverished so that there are 16 parts of air to one of petrol vapour, then it will not ignite at all.

Any deviation from the correct mixture (within the limits of ignitability) will cause the explosions to be weaker, and thus the engine will lose power.

The object of the carburetter is to continually give the engine the first-named mixture, but in practice naturally we only get the nearest approach to this that we can. The best actual mixture on any day varies slightly according to the atmospheric conditions present. (On a fine day more air can be used than on a dull wet one), and in first starting an engine the proper amount is rather a question of guess work. If, on turning the handle the engine does not fire, the mixture will have to be slightly altered until it does. It will be seen, therefore, that it is much easier to get the first ignition with the intensely hot electric spark, than it is with the cooler hot tube. If, from any cause, the oil feed in a paraffin engine, with hot tube, varies ever so slightly from the correct amount, the engine will probably stop, as they are very sensitive about this; but with a petrol engine, with the electric spark, more tricks can be played with the mixture, without the engine actually stopping, though it will always give warning to the practised ear that there is something wrong.

By being able to time the exact moment of ignition, we can not only get the very best possible results out of an engine at any time, but we can, by retarding the ignition, slow down the engine to a great extent, as when coming alongside anywhere, and the engine is always ready to give out its full power by advancing the ignition again to its former position Another advantage is that when starting the engine we need never experience the unpleasantness of a back-stroke, and its consequent jar on the hand.

It must be stated, however, on the other side, that the use of electric ignition introduces a good many complications and sources of breakdowns, and anyone using it should endeavour to master some of the peculiarities of electricity before he goes away for a cruise. In a future chapter the chief sources of trouble will be gone into, and means for correcting them clearly pointed out. If more or less reliability is a *sine quâ non*, then probably some form of hot tube ignition will best insure this. Certain inventors have gone very near to producing a form of hot tube ignition, without any external flame, so that it could be used with safety even with loose petrol about, and the principle on which this was worked was by keeping a small platinum tube in a state of incandescence, by the same means as the platinum points for doing poker work or "pyrography" are kept hot, that is by directing a jet of petrol vapour mixed with air into the interior of the point. So far these experiments do not seem to have reached a practical form, though it is quite possible that they should do this at any moment.

D

Although the hot tube system might be thus rendered quite as safe as the electric system, still the latter will always have one great point in its favour, and that is that with it the engine is always ready to be started in a few seconds of time. This is often of considerable advantage in an auxiliary; and in the author's boat the engine is often brought into use just to enable her to weather a point or shoal, which she would not do under sail alone; whereas if there was any lamp to be lighted and tubes to be heated the engine would not be so used.

As an example of the state of perfection to which the manufacture of the petrol engine has been brought, it is worthy of consideration that many of the present engines can be run up to 2,000 revolutions per minute, and even over! Now an engine running at 1,800 means that each revolution of the crank occupies only one-thirtieth of a second of time, but as each revolution means two strokes of the piston, a stroke has to be done in one-sixtieth of a second! Yet, in this minute fraction of time, the charge of petrol has to be turned into a gas from a liquid form, the inlet valve has to open and close, and the mixture has to be drawn into the cylinder. It is very difficult for the mind to grasp the possibility of this, but it clearly shows how important the weight of each moving part must be, and also what *thousands* of experiments must have been made before the most suitable metal or alloy could be found for it, and how light it could be made to be able to stand the excessive vibration, combined with high pressures, and the presence of flame entailed by use in an internal combustion engine.

The French makers of motor car engines seem to have decided that these excessively high speeds were not obtainable with durability, and the speeds most favoured by them lately seem to range from 750 up to about 1,200 revolutions per minute, so as to save some of the excessive wear and tear. And it must be remembered that the marine motor has a great deal harder task set it than has the land motor, for the latter only has to work at its utmost power going uphill, on the level it only works with a light load, whilst going down hill it gets a holiday, combined with extra cooling. The marine motor is always going uphill, and the screw offers a dead resistance from the moment of starting till the end of the journey. On the other hand, however, the marine motor has a continual stream of cold water circulating through its jacket to keep the cylinder cool, whereas the land motor, after running some miles, has somewhat warmed up its supply of cooling water, and does not therefore get the same effective cooling.

Fig 17 shows a Daimler launch motor, with hot tube ignition, though they are also fitted for electric ignition, and it will be seen that it is a two-cylinder one, the cylinders being side by side, and working on to one common crank, as is shown in Fig 18; by this means, an explosion is obtained at every revolution of the flywheel, as in the case of the two-cycle engine, with one cylinder, so that the flywheel can be kept down in size and weight, and the engine runs more evenly and quietly. The lamps burn in an iron case, closed by doors with perforations, which admit the necessary

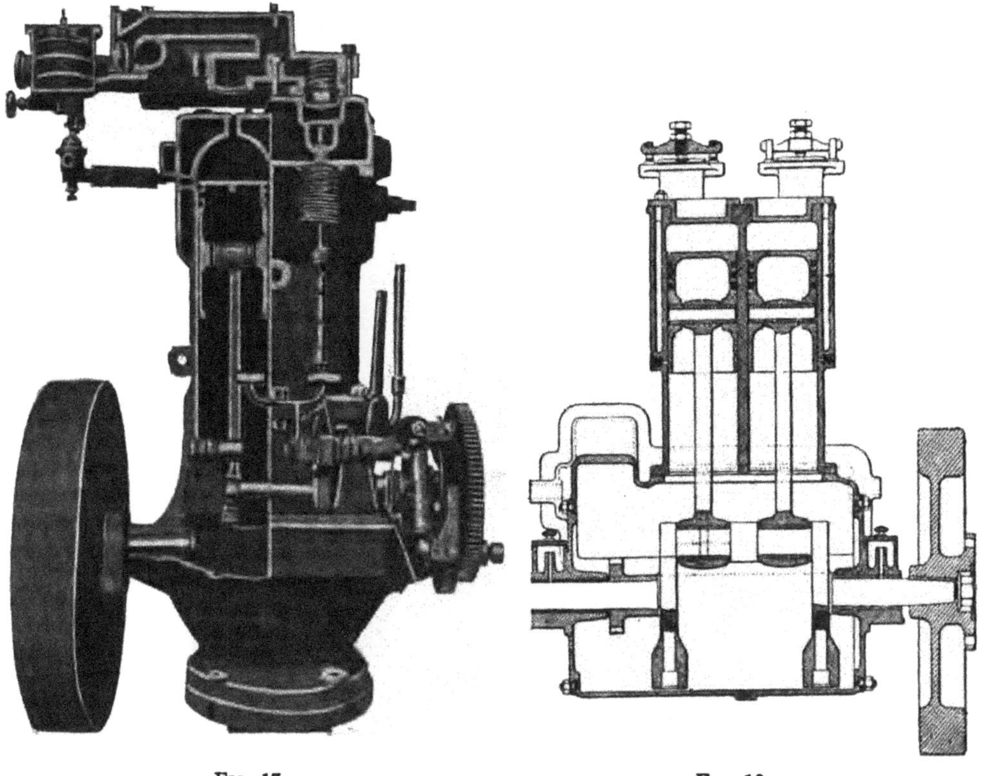

FIG. 17.
DAIMLER ENGINE.

FIG. 18.
DAIMLER TWO-CYLINDER ENGINE.

air, whilst keeping off draughts; the air entering the carburetter is warmed by the lamps first. This is a very essential point, as will be seen when we come to explain carburetters, as, otherwise, even in summer weather, a sort of hoar frost is formed by the constant evaporation, which interferes with the proper working of the carburetter. When the speed of the flywheel reaches a certain amount a governor comes into play, which prevents the exhaust valves from being lifted, so that in this case, no fresh charge can be drawn into the cylinder and fired. As soon as the speed falls to the proper amount, the governor again allows the exhaust valves to be lifted. Both valves are not cut out at once, but first of all one is, and then should the speed not be reduced enough, the other is cut out of gear.

Although, as stated above, the Daimler were the first firm to introduce the petrol engine here, they still hold their own, and turn out excellent work, and launches fitted with their motors are to be constantly seen, both on fresh and salt water.

Fig. 19.

Daimler Launch.

Engines of 3, 6 and 10 h.p. are made with two cylinders each, whilst larger powers of 20, 35 and even 60 h.p. are made with four cylinders each. The 3 h.p. one weighs just under 2 cwt., and the 10 h.p. one just over 2½ cwt. A yacht's boat fitted with an engine is shown in Fig. 19.

Another form of petrol engine is coming generally into use here, and is shown in Figs. 20, 21 and 22. It is known as the *Simms* motor, and is

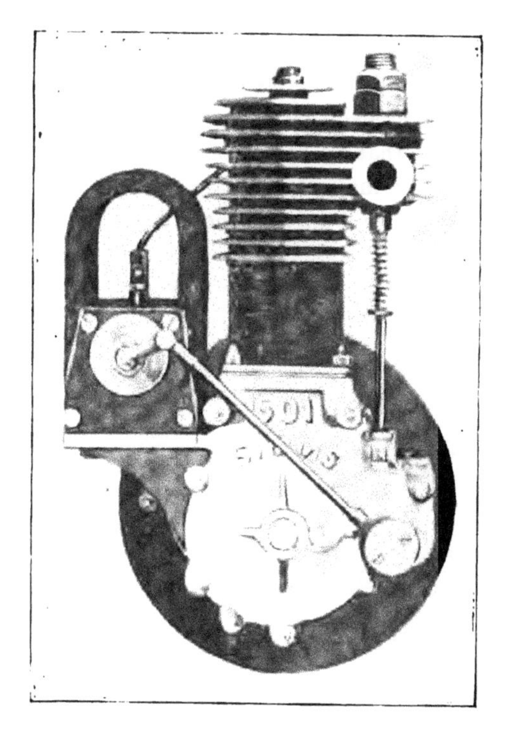

FIG 20.—SIMMS ENGINE, SHOWING MAGNETO.

FIG. 21.—SIMMS TWO-CYLINDER ENGINE.

FIG. 22.—SIMMS ENGINE, WITH CLUTCH AND STARTING GEAR.

made by that company, of Bermondsey, London. It is a very high speed engine, the reciprocating parts being made very light, and its chief feature is the great power obtained, owing to this, and also that it carries on its bedplate the complete ignition gear (electric) so that no batteries, or coils, or outside wires are required. This Simms-Bosch magneto-electric ignition will be fully described in a later chapter, and we will only say here that it

Fig. 23.

Mitcham Four-Cycle Engine.

makes the engine entirely automatic, so that it can be taken out and put into a boat with fewer connections than are required with other engines.

At present the single-cylinder engine is made in three sizes, *i.e.*, 6 b.h.p. running at over 2,000 revolutions per minute and weighing only 142 lbs.; the 3½ b.h.p. running at the same speed and weighing 90 lbs.; and the small 2 b.h.p. also running up to over 2,000 revolutions, and weighing

31 lbs. This small one was to be seen running at the recent Stanley Show, with gas. The weights, as above, include the magneto-electric apparatus.

The two larger sizes are water cooled, but the small one is only air-cooled. They are all fitted with timing gear, by which the revolutions can be controlled from 200 up to the maximum. Another system of regulating the speed is by regulating the lift of the inlet valve. The smaller the lift the less mixture can be drawn into the cylinder at each stroke.

Higher powers are made with twin cylinders, and these give 8 and 12 h.p., whilst the manufacture of still larger sizes is being contemplated.

A very compact form of electric light generator for sailing yachts is brought out by this firm, consisting of an engine and dynamo, all fitted on to a common bedplate, the dynamo being driven by the engine shaft direct, so that no belts or gearing are required.

Fig. 23 shows a four-cycle engine as made by the Mitcham Motor Co. This firm, whilst making a *specialité* of two-cycle motors, also supply the others in order to suit all customers. As shown in the drawing, these engines are made with two and four cylinders, giving powers of 12 and 24 h.p.

As was the case in dealing with paraffin oil engines, the petrol engines described above only represent some of the types now on the market. New firms are constantly entering into this trade, and bringing out some form of petrol motor, and it is, therefore, impossible within the space of these chapters to do more than discuss some of the most divergent types.

Most of the motor car engines are adaptable for use afloat, provided that certain small alterations are carried out, such as the carburetter and the gear for transmitting the power. For use in salt water, however, it is advisable to use only certain metals and connections for the cooling water, so that before buying a land motor this point should be gone into.

In the next chapter we shall discuss some of the leading two-cycle engines.

CHAPTER V.

SUMMARY.

TWO-CYCLE SYSTEM FURTHER DESCRIBED AND EXPLAINED—LOZIER—MITCHAM POPULAR — SIMPLEX — ROCHESTER — MONARCH — AND TRUSCOTT PETROL TWO-CYCLE ENGINES, AND THEIR PECULIAR FEATURES.

We have now to consider and explain the two-cycle engine, which, as explained in a previous chapter, differs essentially from the four-cycle engine in that it has an explosion and working stroke of each down stroke of the piston. By working on this system, it is enabled to be made with fewer parts, and greater simplicity, and these two advantages render it very suitable for certain forms of marine work, such as small launches or auxiliaries.

In order to explain how the cycle works, we cannot do better than make use of some of the drawings which appear in the book of the Lozier Co., which is one of the leading motor manufacturing companies in America, and no one on seeing this book can fail to admire the way in which it is got up. The diagrams are of the clearest, and the general principles on which the engine works, even down to the smallest details (which are so often guarded as secrets), are here fully described and given to the public. In fact the book is a work of art, which, however, in the words of the issuing firm, "talks business from cover to cover."

In Fig. 24 we have a section of the two-cycle motor, in which the electric spark has just fired the mixture, and the piston is being driven down, doing, therefore, a working stroke. The next charge to be fired is all ready in the crank chamber under the piston, so that as it descends it slightly compresses this charge.

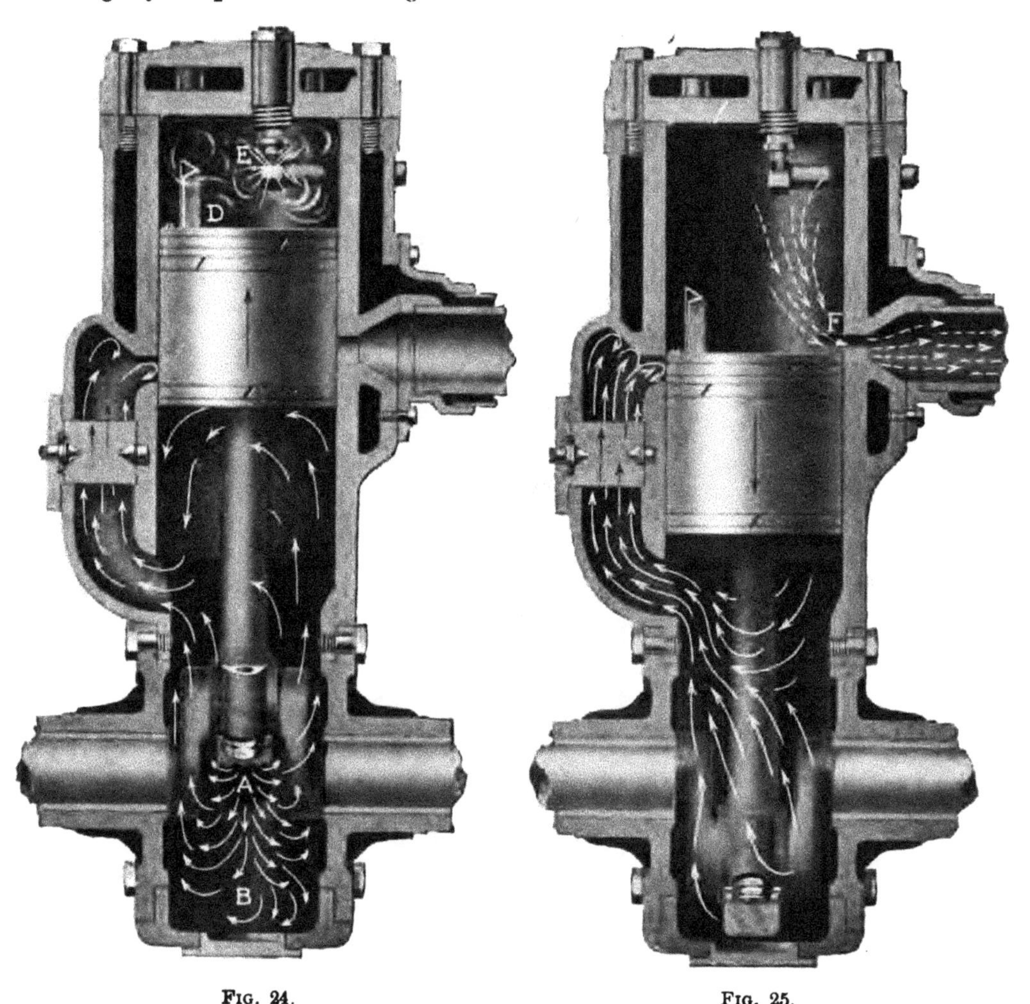

FIG. 24.
TWO-CYCLE ENGINE—FIRING.

FIG. 25.
TWO-CYCLE ENGINE—EXHAUST.

Just before the piston reaches the bottom stroke, it uncovers first of all the opening in the cylinder F, Fig. 25, which communicates with the exhaust, and the burnt gases therefore escape through the exhaust drum into the air. The piston now descends a little more, and the entry port on the opposite side, C, Fig. 26, which communicates with the crank chamber, is now uncovered, and the fresh charge, which is slightly compressed, enters the cylinder, driving out the greater part of the burnt gases of the previous charge. Just opposite the entry port is a deflector fixed on the

FIG. 26.

TWO-CYCLE ENGINE—ADMISSION OF FRESH CHARGE.

top of the piston, and the action of this is to deflect the new charge up to the top of the cylinder, so as to be *above* the already burnt gases. If it were not for this, the new charge would have a tendency to enter at one side of the cylinder and go out at the other. We must now imagine the piston to be rising again, and, as it does so, it first of all closes the entry port, C, and then the exhaust, F. On further rising it now compresses the charge above it, and at the same time draws a fresh charge of mixture from the carburetter into the crank chamber, where it is well churned up and mixed by the action of the revolving crank. Just before the piston reaches its top stroke, the charge is fired by the electric spark, and the cycle is repeated.

It will now be seen that we here get rid of the usual inlet and exhaust valves, together with the gear actuating the latter, and we only have one non-return valve between the carburetter and the crank chamber, and this is away from all flame and heat, and is, moreover, only subject to a slight compression. This is what renders the two-cycle engine so much simpler, but its proper working depends almost entirely on the design. It has taken years of experiments to find out just the best position, size and shape of the two ports in the walls of the cylinder, and also of the deflecter. The parts projecting into the cylinder, which form the electric ignition, also have to be made of a certain size and shape (independently of any electric requirements), as otherwise, owing to the fact that the explosions follow each other regularly, any such projecting part is very liable to get so hot that it would fire the charge as soon as it entered the cylinder, and thus produce a *back stroke*. In the four-cycle engine there is a cooling stroke between each explosion which prevents any such part becoming overheated. We have emphasised the necessity of accurate design in the two-cycle engine because it explains somewhat how it is that many firms have given up their manufacture, and have gone back to the four-cycle system. On the other hand, it reflects the greater credit on those firms who have succeeded in turning out the reliable engines that are now to be obtained working on the two-cycle system.

It will be noticed that in two-cycle engines it is a very important matter for the crank case to be quite airtight, and for the piston rings to be a very good fit in the cylinder, because, if the crank case admitted any outside air on the suction stroke, it would not only tend to weaken the mixture, but would also let out some of it on the down or compression stroke instead of forcing it in properly into the cylinder, and if the piston rings let any of the fired gas past them this would get into the next charge and spoil the force of its explosion, even if it did not actually fire it. The bearings of the crank shaft should be of a good length, as well as a good fit, and it is better to lubricate them with a stiff grease from a pressure cup than with the usual oil, which gets liquid when heated.

It will also be noticed that whereas in the four-cycle engine the fresh charge has the time of a whole stroke in which to enter the cylinder, and the exhaust has a little more time than this in which to escape; in the two-cycle engine, both the exhaust and the entry have to be made in a *fraction* of a stroke, and the consequence is that the latter engines cannot be run at the same speed as the former. Moreover, as there is always a certain portion of the old charge left in the cylinder under the new charge, this has a tendency to render the explosion weaker than it would be in the same cylinder with a four-cycle system. This explains why we do not get the same power out of the cylinder, but in many cases this loss of power is more than counterbalanced by the simplicity gained.

A well-known American expert gives the power of the two-cycle engine, as compared with a four-cycle one with the same cylinder, and running at the same speed, as only 50 per cent. greater. If the explosions were equal in force in both cases, the two-cycle engine would give double the power, as it has twice the number of explosions, so that we may take it that each explosion of the two-cycle engine gives about three-quarters of the power of the four-cycle one.

One peculiarity of most of the two-cycle engines is that whichever way they are started, they will continue to run. In fact, by intelligent manipulation of the electric spark, some makes of engines can be slowed down, and just before stopping they can be reversed, so as to come alongside anywhere. If properly designed, they should not be liable to reverse themselves when running, unless the mixture is too poor in petrol, and consequently weak, or if the revolutions are suddenly checked by the screw fouling thick weeds, or a rope for instance, when this may perhaps happen, but, in the two latter cases, the reversal of running tends to clear the obstruction, and can be immediately rectified by the handling of the electric sparking gear.

Two-cycle engines are rather more sensitive about their mixture than four-cycle ones; thus, if the mixture is a poor one, the combustion in the cylinder is so delayed, that when the fresh charge enters from the crank case it meets the flame of the still burning former charge; this ignites the fresh charge down in the crank chamber, and the next stroke (or possibly two strokes) is a useless one. Whenever an explosion is heard in the crank chamber, more petrol must be given, or the engine will be liable to stop.

In the passage leading from the crank chamber to the cylinder is fitted a throttle valve, by means of which the engine can be slowed down for as long as is wanted. In some engines a screen of wire gauze is also inserted, chiefly to keep the lubricating oil in the crank chamber from being splashed up so as to be drawn into the cylinder together with the charge.

With these general remarks on two-cycle engines, we will now follow the Lozier engine, which possesses some interesting differences from other types. An elevation of this engine is shown in Fig. 27.

Fig. 27.
Lozier Engine.

Lubrication.—This very important detail is carried out, not only by keeping a certain quantity of lubricating oil in the crank chamber, where it is splashed about by the dipper attached to the lower part of the crank brass, and shown in Figs. 28 and 29, but a very ingenious system is used in order to constantly supply a small portion of oil all round the piston, which is just where it is wanted.

In Fig. 30, a sight-feed oil cup is set to give so many drops per minute, and these drops fall down by gravity past the little valve E into the channel communicating with the cylinder, and right through the water-

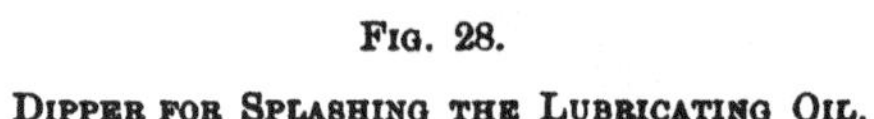
FIG. 28.

DIPPER FOR SPLASHING THE LUBRICATING OIL.

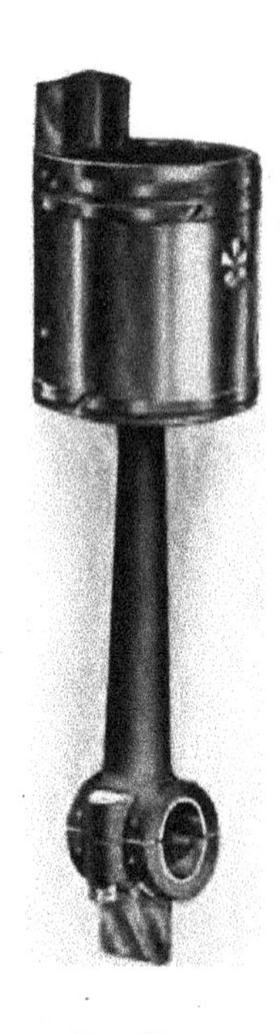
FIG. 29

THE PISTON.

jacket. Just before the piston reaches its bottom stroke, it uncovers the small hole in the walls of the cylinder B, and a portion of the burnt charge rushes down into the annular groove right round the piston, between its two upper rings and it carries with it the drops of oil which have there accumulated; the piston therefore constantly has a ring of lubricating oil swept round it, and from thence the oil gets properly distributed evenly on the piston rings.

Each charge is carburetted by a spraying valve, which is of the simplest form, the amount of petrol being regulated by a graduated screw valve, and the air is warmed before going to the petrol by drawing it from a jacket surrounding the hot exhaust pipe. This spraying valve will be fully described in the chapter dealing with carburetters.

The exhaust gas after leaving the cylinder goes through a cold water jacketted pipe, which reduces its volume considerably, and the exhaust drum is also similarly cooled, so that it can never get hot enough to char any woodwork, which is often the case in unjacketed exhaust drums not properly fitted into a boat.

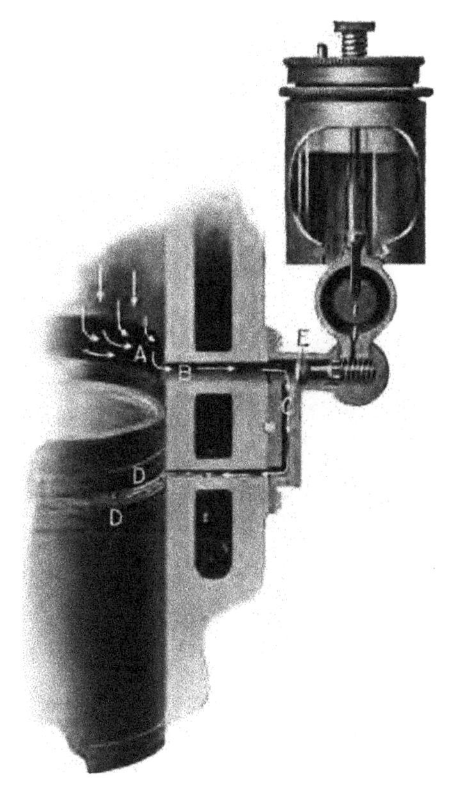

Fig. 30.
Lubricating the Piston.

The circulating pump is of the simplest description, as used in the De Dion Motor Cars, and is driven off the main shaft by a chain like a bicycle one.

One of the most important adjustments of all internal combustion engines is that of the mixture of air and the gas used for the particular engine. The best mixture is generally found by slightly varying the mixture valve and listening to the beat of the engine, but for those whose ears are not sufficiently trained for this the eye can be used by noting the colour of the flame of the exhaust gas as it issues from the cylinder. This, of course, is impossible in most engines, but in the Lozier and kindred engines a small relief cock is fitted into the combustion chamber, and by opening this, when the engine is running, the colour of the issuing flame can be observed, and this should always be of a deep blue colour, with no signs of yellow or white about it, which would indicate too much petrol. A very pale blue flame would indicate a too poor mixture.

The electric ignition is of the form giving the most reliable results, and will be fully described in the chapter dealing with ignition. We will only say here that dry batteries are used to start the engine, but when once

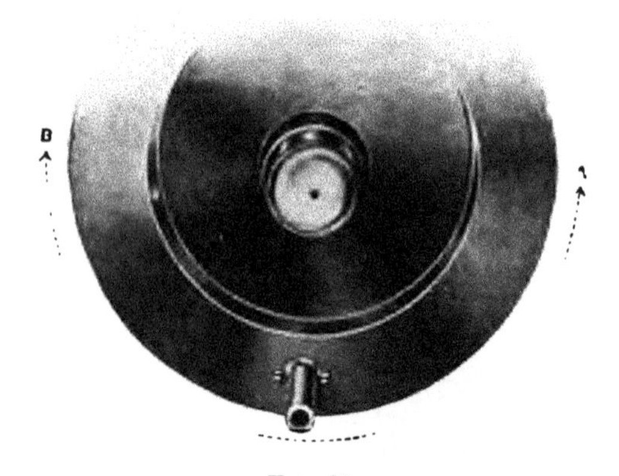

Fig. 31.

Starting a Two-Cycle Engine.

started it generates the requisite supply of electricity by means of a small dynamo driven off the flywheel.

In the general remarks on two-cycle engines, in a previous chapter, it was stated that they were easier to start than are the four-cycle ones; and the reason for this may be shown in Fig. 31. In four-cycle engines the flywheel has to be turned round completely for more than one revolution, in order to introduce the necessary charge, then to compress it, and lastly to fire it. If the mixture is not exactly right, this operation may have to be repeated several times, and, although the full compression is not felt (owing to a relief cock being opened when starting) still a good deal of muscular force is required. In a two-cycle engine, however, the flywheel has only to be pumped backwards and forwards, between the positions A and B, several times, and this draws the necessary mixture into the crank chamber and cylinder. By then giving the flywheel an extra push, so as to get the handle into the position C, the electric spark will pass and fire the mixture, and the engine is off. In most of the two-cycle engines, if the handle were brought into a position on the other side corresponding to C, the engine would start in the opposite direction, and continue running thus. In many engines the handle sticking out of the flywheel is a constant source of

FIG. 32.
BALL-BEARING THRUST.

danger to anyone near it when the engine is running, but in the Lozier, directly it is let go, it flies back into the flywheel, by the action of a spring, and remains there until wanted again for a fresh start.

A reversible bladed propeller is fitted for these engines, the thrust being taken by a ball bearing, A, Fig. 32, so as to minimise friction.

The engines are made at present of four sizes with single cylinder, *i.e.*, 1¼, 2½, 4 and 6 h.p. (the 1½ h.p. weighing 230 lbs. and the 6 h.p. 795 lbs.), whilst the higher powers of 8, 12, 16 and 20 h.p. are made with two cylinders (the 8 h.p. weighing 800 lbs., and the 20 h.p. 1,650 lbs.).

Messrs. F. R. Leith & Co., 54, Piccadilly, London, W., are the sole agents for these engines for Great Britain and Ireland, and will give any information concerning them.

Fig. 33 shows a two-cycle engine made by the Mitcham Motor Co., of Mitcham, Surrey. These engines work on exactly the same principle as the Lozier, and only differ from them in certain small details. The same sort of electric ignition is used, and the current can be supplied either by dry batteries or by a small dynamo driven by the engine itself.

The engines are made in five sizes, *i.e.*, 1½, 2, 3, 5 and 7 h.p., the weight of the smallest being 140 lbs., and that of the largest 650 lbs.

Fig. 33.—Mitcham Engine.

A reversible bladed propeller, with spare blades which can easily be inserted in case of accident, is supplied with the engine.

Another engine is the new Popular, a petrol motor sold by Messrs. Lister, of Harrow Road, London, shown in Fig. 34, and it is also a two-cycle one. It is a well-designed engine, and the spraying valve, or carburetter is attached to the crank chamber, the petrol being led down to it there. The electric spark can be retarded or advanced whilst the engine is running so as to get the best results, and a throttle provides a means of slowing down the engine still more when required. Cylinder lubrication is obtained by a drip lubricator, and when once the proper drip feed has been found out and adjusted by the regulator, no further adjustment of this important detail is required, as the oil feed is cut off by a separate cock whenever the engine is stopped.

Fig. 34.—Popular Engine.

Some very valuable information, together with full instructions for fitting, adjusting and working the Popular motors is given by Messrs. Lister to customers. The engines are supplied in four sizes, *i.e.*, 2¾, 4, 8 and 16 h.p., the former weighing 165 lbs. and the latter 1,250 lbs. Complete motor boats are also supplied either of the small open type or of a larger and decked type.

Special electric batteries are used for giving the spark at starting, and as soon as the engine is running well a magneto driven by the flywheel gives the current and saves the batteries.

Another type of two-cycle engine is the " Simplex," made in America, and which can be seen at the Company's offices at 120, Southwark Street, London. It possesses one of two peculiarities, and the most notable one is that the petrol is vaporised by admitting to it air which has been considerably warmed by a heater through which the exhaust passes ; the effect

FIG. 35.—SIMPLEX ENGINE.

of this is that the engine can use spirit of a much heavier nature than the ordinary petrol, and, in fact, will work well with benzoline and other similar spirits. The electrical igniter is of an efficient and simple pattern, with platinum points, and the whole of it can be easily removed from the cylinder for cleaning. Fig. 35 shows the engine, with the heater at the side of the cylinder. The screw is reversed in direction by a neat clutch attached to the engine bed-plate, and the engine can be run free from the screw at will. Single cylinder engines are made of 2 to 10 h.p. (the former weighing 450 lbs., and the latter 1,225 lbs.), for larger sizes double cylinder engines are made of from 5 to 20 h.p. (the former weighing 800 lbs., and the latter 2,400 lbs.).

Another two-cycle engine possessing some peculiar features is the "Rochester," of American make, and sold by Mr. Alfred Burgoine, of Kingston and Hampton Wick. Fig. 36 shows one of these engines of

FIG. 36.

ROCHESTER ENGINE.

2 h.p. In these engines the mixture is drawn into the crank chamber through a special carburetter, and on the down stroke it is forced up through an inlet valve, thus not only keeping the sparking points cool but by getting the fresh charge in at the *top* of the cylinder it tends to keep it separate from the old burnt charge, which is forced out through an exhaust

in the side of the cylinder, as in the two-cycle engines described above. By regulating the movement of the inlet valve, the speed of the engine is also regulated, and the timing of the electric spark forms also another form of speed regulation. A very neat form of reversing clutch permits the screw either to be disconnected from the engine, or to be reversed in direction for going astern. They are made in sizes from 1 h.p. up to 18 h.p. and have one, two or three cylinders, according to the size. The 1 h.p. engine weighs 125 lbs. and the 18 h.p. 1,800 lbs. approximately.

Another two-cycle engine is the "Monarch," sold by the Monarch Motor Co., of Eastcheap, London, and it possesses one or two points differing from the engines described above. One point is that the electric ignition (although of the same electrical nature as that employed in the engines above) admits of being timed in three gradations; thus the late spark for starting can be used, and as the engine runs at increasing speed the spark can be advanced so as to get the increased power thus obtainable. Greater efficiency is also claimed for the way in which the petrol is admitted to the crank chamber, and there vaporised drop by drop, and the separate valve admitting the petrol has the advantage of not being liable either to clog up with some of the waxy deposits often produced by the evaporation, or to leak when the engine stops running.

A reversible bladed propeller, with interchangeable blades, is supplied for the Monarch Motor, and a governor prevents the engine from racing when relieved of the weight of the screw. The engines are made in nine sizes with single cylinder, *i.e.*, ¾, 1, 2, 3, 4, 6, 8, 10 and 15 h.p., the smallest weighing 100 lbs. and the largest 1,600 lbs. Two-cylinder engines are made in six sizes, *i.e.*, 4, 8, 12, 16, 20 and 30 h.p., the smallest weighing 400 lbs. and the largest 2,600 lbs. The three cylinder engines are made of six sizes, and range from 6 to 45 h.p., *i.e.*, 6, 12, 18, 24, 30 and 45 h.p., and the smallest weighs 600 lbs. and the largest 3,600 lbs.

Fig. 37 shows the Truscott two-cycle petrol motor, which is gradually becoming known and used here; several boats now building, and existing ones going in for auxiliary power, are utlising this motor. It is made in two styles, *i.e.*, the Special and the Standard. The Special is rather lighter than the other, in most sizes, and it has the usual spraying valve attached to the crank chamber, and the petrol is led down to the valve, and there sprayed and vaporised as it is wanted for each stroke. The Standard draws its vapour from a combined tank and surface carburetter in the bow of the boat, and it is claimed that by this means a spirit of lower specific gravity than the usual petrol, such as benzoline for example, can be used whilst giving full power with the engine. There are several destinctive features in the design of the engine and its accessories, such as long-bearing surfaces, a good system of lubrication, readily accessible pump and ignition gear; and speed regulation is obtained by a throttle, and also by a timing device for the ignition, which is of the low tension electric system. Batteries are used to start the engine, and are then switched off, as the dynamo comes into action. The starting handle is so arranged that it comes free from the flywheel altogether as soon as the engine starts. The brake horse power given in the catalogue refers to the power given off *at*

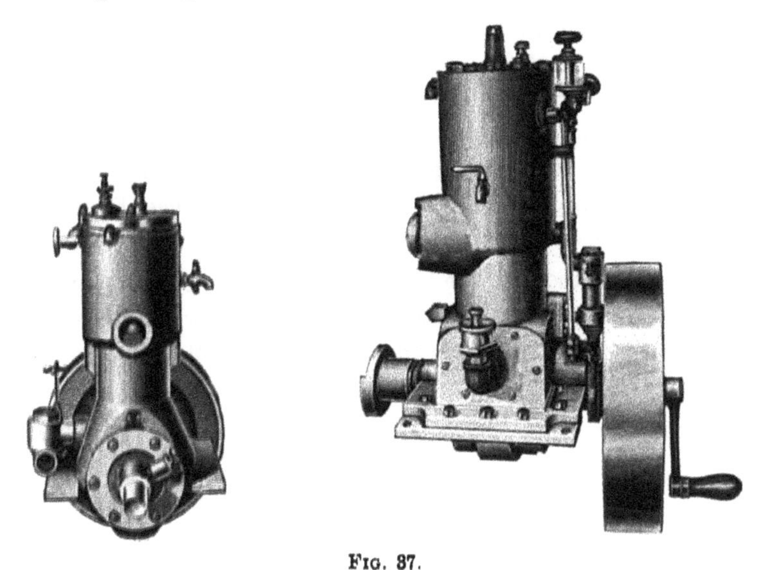

Fig. 37.

Special Standard Truscott Engines.

the propeller. Messrs. George Neill, of 38, Fenchurch Street, London, are the sole agents for this motor, which is made in America, and catalogues, both of the engines and also of boats and accessories, can be obtained from the agents.

The Special motors are made with one, two, or three cylinders, and vary in power from 1½ h.p. up to 25 h.p., the weight of the smallest being 165 lbs. and that of the largest 1,060 lbs.

The Standard motors are made with one or two cylinders, and vary in power from 1½ h.p. up to 17 h.p., the weight of the smallest being 155 lbs. and that of the largest 975 lbs.

The base of the Standard motors is arranged so that the bed on which the engine rests in a boat is in line with the shaft, and not below it as usual, and it is claimed that by this means the vibration is much reduced.

The timing gear for the electric ignition is so arranged as to be automatic, *i.e.*, the late spark for starting, and as soon as the engine begins to run well the spark becomes automatically advanced. This is done by fitting the cam on the engine shaft, which works the pump and the sparking points, not as usual tightly keyed on, but free to work backwards or forwards through a certain angle.

This cam is so held, by two spiral springs on the flywheel, that its usual position gives the late spark, and with this spark the engine is started by hand. As soon, however, as the engine begins to run two hinged arms on the flywheel begin to open outwards, owing to the centrifugal force acting on weights at their outer ends, and as these arms open thus outwards they turn the cam through a small angle, so as to advance the spark just so as to get the best results for a given speed.

We have now described the leading types of two-cycle engines, and the next chapter will be devoted to a description of the various carburetters and vaporisers used both by two-cycle and four-cycle petrol motors.

CHAPTER VI.

SUMMARY.

CARBURETTERS—VARIOUS WAYS OF MAKING AND MIXING THE CHARGES FOR ENGINES—LONGUEMARE—LOZIER—AND FORMAN CARBURETTERS, AND THEIR PECULIAR FEATURES.

ALL the engines that we have described in the previous chapters are "gas engines," that is, they would all work well if connected on to the town mains of a gas supply. As, however, they have to work with a portable fluid, this fluid has to be rapidly and economically turned into a gas, as wanted for the engine, and this is the province of the carburetter.

It was previously stated that the best mixture for the engine was composed of one part of petrol vapour to about 8·3 parts of air. This is about equivalent to 39 grains of petrol per cubic foot of air, so that 1 fluid ounce measure of petrol will carburet about 8 cubic feet of air, or about 8,000 times its own volume! It will be understood, therefore, how very *exact* the action of the carburetter must be (especially those working on the jet system) in order to get anything like the proper mixture in a fast-running engine. Owing to the very high speed at which oil engines work, each suction stroke has not time to fill the cylinder, so that if we calculate the number of cubic feet of air drawn in theoretically in a given time, we shall find that a fluid ounce measure of petrol will carburet about 10, 12 or even more cubic feet of air as calculated; but this is because the engine really draws in a lesser quantity of air in working, as explained above.

Carburetters may be roughly divided into two classes, *i.e.*, those working on the "surface carburetting system," and those working on the "jet," or "spray" system. The latter are called *vaporisers* in America, and are also known often as *atomisers* or *spraying valves*. The function of all of them, however, is to *carburet* the air before it goes to the cylinder, so we shall here designate them all carburetters, mentioning, at the same time, the principle on which they work. Taking the surface carburetters first, we may say that they act by drawing the supply of heated air either through or over the surface of the petrol, or through a system of lamp wicks, or something of a similar nature, which will present a great surface of petrol to the incoming air.

Fig. 38 will explain the action of the surface carburetter more clearly. It consists of a containing vessel for the petrol, and on the surface of the liquid floats the ring of cork or spun metal supporting the tube AB, which

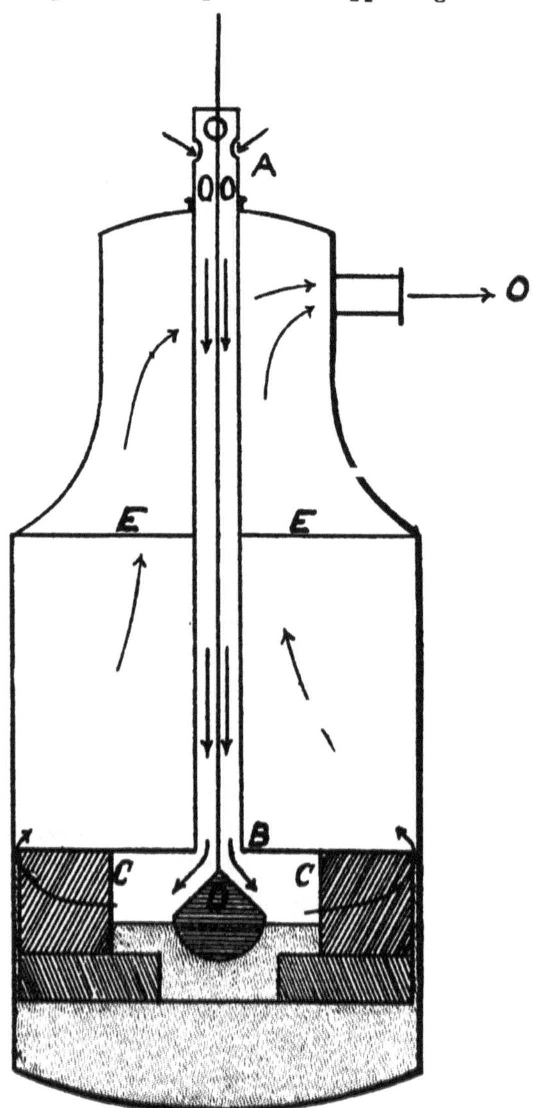

FIG. 38.—SURFACE CARBURETTER.

is free to slide up or down through the lid of the vessel. The outlet O is connected with the engine, and, as it draws in its mixture, it creates a partial vacuum in the vessel, and the outside air rushes in down the tube

AB, entering through the holes in its upper end. The incoming air now strikes against the conical float D attached to a wire passing up through the tube AB. The conical float divides the current of air all round so that it strikes against the surface of the petrol, and passing through slits in the floating ring passes up round the edge of the metal disc soldered to the tube AB, and rises up as a mixture of air and gas. It now passes through a metal diaphragm EE, perforated with numerous small holes, so as to further

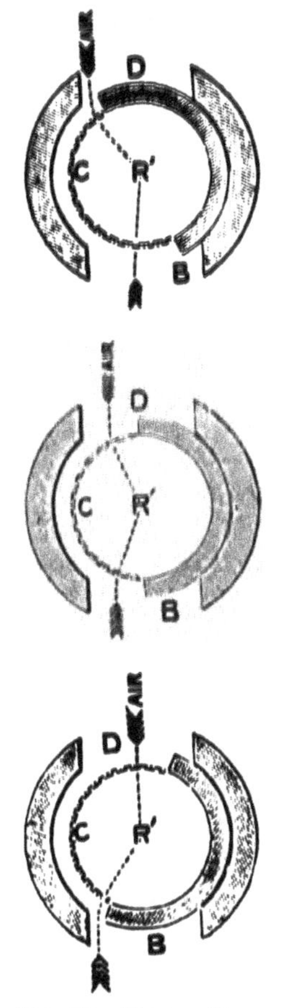

FIG. 39.—MIXING VALVE.

mix the air and gas, and then goes on to the engine through the outlet O.

As the level of the petrol gradually sinks, the float and the pipe AB also fall so as to keep the same position as regards the liquid.

As the mixture is nearly always too rich for the engine, it has to be further diluted, and this is done through the mixing valve, Fig. 39, where R is the pipe leading to the engine, and the opening B is that connected to the carburetter, and through which the mixture comes. D is an opening communicating with the outside air. The dotted line C shows a screen of wire gauze. It will be seen that with the position of the valve shown in the upper diagram, the engine is getting a very small portion of air as compared to the mixture ; whereas, when the valve is turned by hand, as in the centre diagram, it is getting equal parts of each ; and in the lower figure, when the valve is turned still more, the engine only gets a very small portion of mixture to a large one of air.

According to the richness of the mixture obtained at any time from the carburetter, the proper amount of air can be mixed with it before it enters the cylinder, so as to give the best results.

The great drawback of this system is, that inasmuch as petrol is a complex mixture of various spirits with different evaporative qualities, those which are most volatile go off first, and make a capital gas for the engine, but the consequence is that, after running some time, only the heavier constituents of the petrol are left, which require the air to be more heated in order to evaporate them quickly enough, and should the engine be stopped, and cool down, then, on trying to start it again, great difficulty would be experienced. The consequence is, that in the Daimler engines, which use to use surface carburetters, the stale petrol left in the container had to be thrown away, and fresh petrol put in for each new start.

Another drawback was that at first the mixture of petrol vapour and air coming from the carburetter would be very rich, so that the mixing valve would be set to give a very little gas and a good deal of air, but as the gas became weaker the mixing valve would have to be continually altered so as to give more and more gas and less and less air, so that constant attendance would be required ; and a further drawback existed from the fact that the flame in the cylinder of an oil engine may possibly get back into the carburetter and thus cause an explosion, which might be serious should the carburetter tank contain any large amount of petrol. It should be noted that as long as the gas from the carburetter is coming away so rich that it requires a good deal of air to be mixed with it in order to be fired in the engine, there is no chance of a flame running back into the carburetter any more than it could run down a gas pipe into the mains of a town. It is when the gas is coming away so weak that it requires no air at all to be mixed with it for the engine that this risk exists. The chances of "back firing" into the carburetter are very small, but it may happen at any time if the inlet valve either breaks or does not close properly at the firing time. There is, of course, just the same chance of the flame getting back into the jet carburetter, but in this case no harm would be done, as there would be only a drop or two of petrol to ignite, and this would burn out almost instantaneously.

It is customary to guard against this back firing by inserting in the pipe leading from the carburetter to the engine, a series of discs of metal gauze, and the gauze should be of a fine mesh, not less than 40 to the inch, and moreover five or six discs should be used, and they should be slightly separated from each other. They are intended to work on the principle of

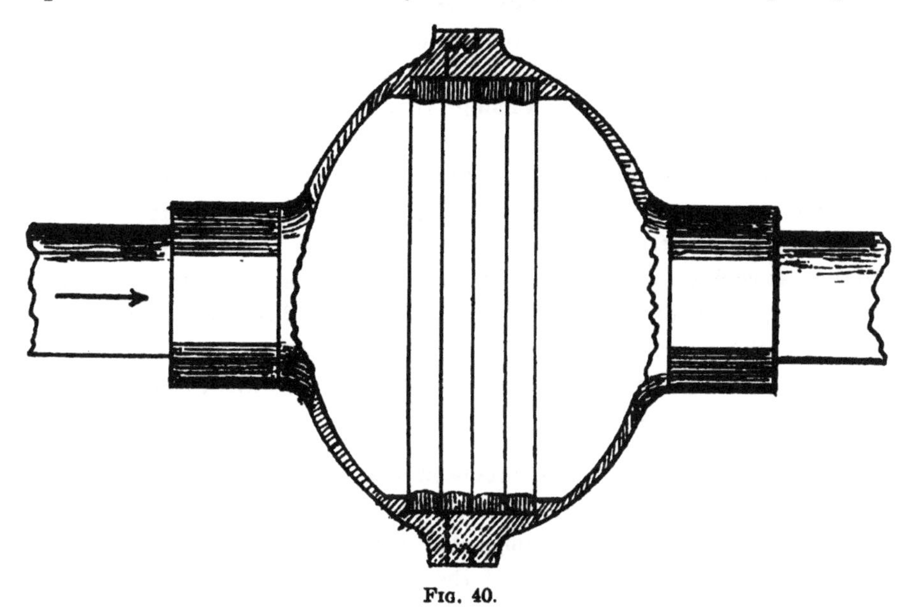

FIG. 40.

FIREPROOF WIRE GAUZE DIAPHRAGMS.

the Davy safety lamp, which is still used in collieries. Fig. 40 shows one of the gauze screens complete as fitted to the pipe between carburetter and engine. The central part is shown in section; it consists of two dome-shaped brass cups, which can be screwed together, and before doing so, a series of gauze wire discs (five in the drawing) with lead washers between them are placed in position one above the other. On then tightly screwing the brass cups together the gauze discs become embedded in the lead washers, and thus an air-tight joint is made all round them, so that the gas must pass *through* all the discs. The diameter of the discs is made much larger than that of the pipe, so that there shall be no obstruction to the free passage of the gas.

The surface carburetter has a great advantage in its favour as being very reliable, though extravagant; whereas in the jet carburetters now about to be described, as their correct working depends on a very small hole, or slit, being kept free and open, they are at times liable to give trouble, and thus add another source of anxiety to the oil engine. Another great advantage of the surface carburetter lies in the fact that its use does not throttle the supply of gas coming to the engine as much as a jet carburetter does. The action of the latter *depends* on the vacuum produced in it by the suction stroke of the engine, and, in some cases this vacuum has to be artificially increased by closing some of the holes by which the air can enter. Now the greater the vacuum thus produced, the less mixture can the engine get at each stroke, and when running at high speeds, considerable power may be thus lost. Another point is that all engines, when running slow, require more petrol per stroke than when running fast, because in the latter case the cylinder has not time to fill itself with mixture at each stroke, but only gets a reduced quantity. In first starting an engine, the float has generally to be pressed down a bit by a handle fitted for the purpose, so as to cause an extra amount of petrol to flow through, all ready for the first few suction strokes, which, being slow, will only create a small vacuum.

As the greater the vacuum produced in a jet carburetter the greater the charge of petrol, we see that when an engine is running very fast it gets *less* air and *more* petrol (although the supply of petrol may be slightly reduced by the shortened time of the suction stroke), and when running slower the contrary is the case, so that it only gets the *exact amount* when running at the average intermediate speed for which the carburetter is set. This is a fault possessed by nearly all jet carburetters, and which has to be corrected by periodical regulation by hand, but they are in general nse, because the drawbacks of the surface carburetter, as described above, are considered to overbalance its advantages. The author, however, has for the last two years made use of a surface carburetter made on the patent *Lanchester* system, which almost entirely does away with the drawbacks mentioned above, and it has proved itself very reliable for use afloat. As the firm of Messrs. Lanchester and Co., of Birmingham, are about to bring out a special form of marine oil engine which will utilise this carburetter, we cannot say any more about it here.

From want of anything better, the surface carburetter held its own until the "Longuemare" jet carburetter came out, and very soon displaced the surface carburetter for all motor-car work.

This carburetter is shewn in Figs. 41 and 42, and it will be seen that its action depends on the petrol being kept at a constant level inside the

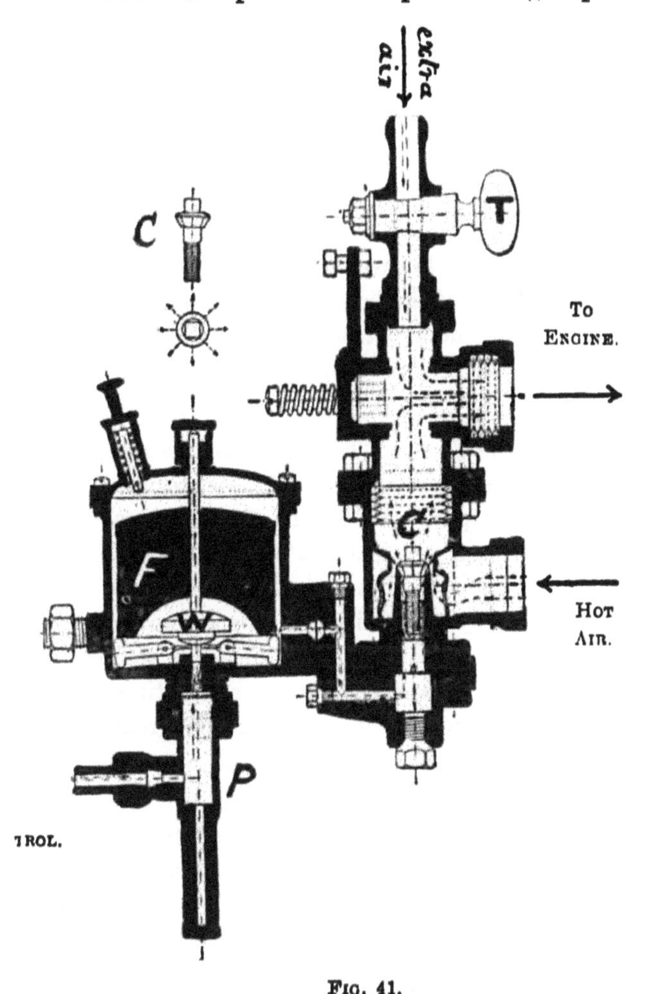

FIG. 41.
SECTION OF LONGUEMARE CARBURETTER.

FIG. 42.
LONGUEMARE CARBURETTER.

little vertical pipe, at the upper end of which is the spraying cone C, and this level must be just below its top. The method by which this level is maintained will be seen by noticing that the petrol arrives from the main supply by the pipe P. and thence runs into the cylinder which contains a cylindrical float F, which almost fills it up, and as the petrol rises it raises up the float, so that it no longer presses down the two hinged levers, and they, in rising, allow the weight W to press down the spindle, so that its pointed end closes the pipe and thus stops the supply of petrol. Of course, the weight of the float is so arranged that it just shuts off this supply when the level is correct. When the engine draws in its supply of air at each stroke, it creates a partial vacuum in the carburetter and in the funnel which embraces the little vertical pipe J, and the consequence is that at each stroke tiny jets of petrol are sucked out through the small orifices in the cone C, and, mingling with the air, form the proper mixture, as it is drawn into the cylinder to be fired.

As the engine draws away the petrol, its level slightly sinks, and the float also sinking, lightly rests on the ends of the levers, which thus slightly open the petrol valve, and admit a little more petrol to the float chamber: and thus the lever is kept practically at a constant height.

As the size in the orifices in the cone C cannot be varied, the method of varying the amount of petrol for each stroke, so as to get the best results, is by varying the amount of the vacuum produced by the suction stroke. This is done by slightly turning the tap T, so that the air outside rushes in, and consequently weakens the vacuum so that not so much petrol is drawn out of the cone C; and moreover, by the addition of this external air, the mixture is still further weakened. As a consequence, therefore, we get a very sensitive form of regulation by this means, and we can at any time be certain that the engine is getting just the correct mixture for the greatest efficiency.

The Longuemare carburetter may be said to be the father of nearly all the others which have since appeared, and their name is legion; a new one used to appear about once a week, and the patent fees must have been quite a source of income to the Patent Office. Now, although this float feed system gives excellent results for land purposes, it does not do so well for sea work, because the motion of a vessel, and particularly a steady list, as in a sailing vessel, upsets the regularity of feed which is so essential to the proper working of an oil engine. The inventors then turned their attention to bringing out a form of jet carburetter which did not depend on a float for its level. One of the simplest of these was shown in a previous article, as that used in the Seal engine, and a modification of this principle is that shown in Fig. 43, and is that used in the Lozier engine. At each suction stroke the air valve B is lifted off its seat, and as it does so, it uncovers a very small hole in the seating which communicates with the petrol pipe F (called gasoline in America). Each jet of air therefore draws in with it a tiny jet of petrol, and thus the proper mixture is made. The amount of petrol entering is regulated by the screw valve D turned by a handle working over a graduated arc, so that the proper position may be always obtained again when once found. When the air valve B is down

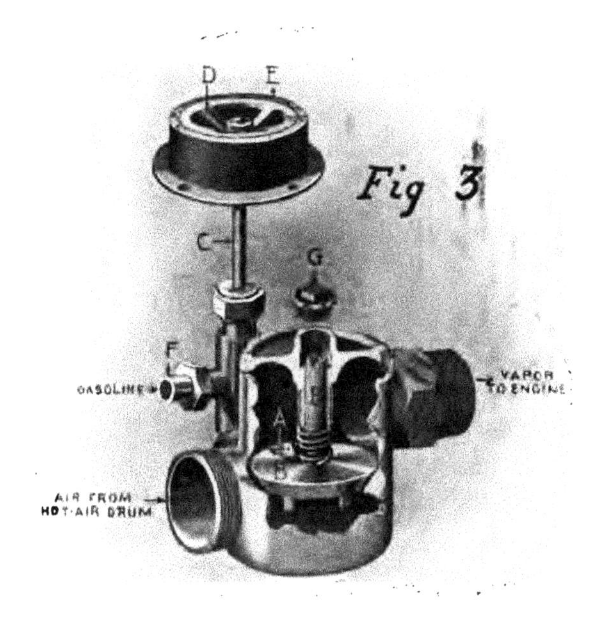

Fig. 43.

Lozier Carburetting Valve.

on its seat, it covers over and closes the petrol hole A. This is, of course, a very simple carburetter, and one which is more or less independent of the motion of the boat, but the delicate part is the tiny hole in the seating of the valve. The constant evaporation of petrol has a tendency sometimes to leave a very fine deposit of a waxy nature, which is quite sufficient to interfere with the flow of petrol, and thus stop the engine. When this happens, screw plug G at the top of the case is taken off, and a screwdriver put down by which the valve is turned round and round on its seat, so as to rub off the wax, and open the petrol hole.

Another form of jet carburetter, independent of a float feed, is shown in Fig. 44, and here the petrol rises up through the small valve A, which is lifted by the suction of the engine, so that there is not so much chance of its getting choked, either by wax or any foreign substance finding its way

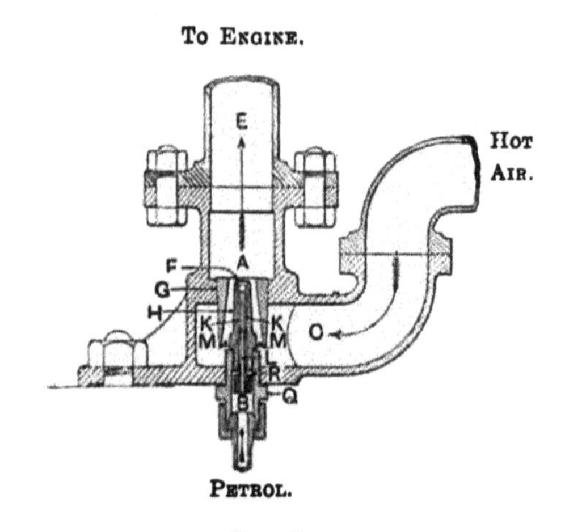

Fig. 44.

Forman Carburetter.

in with the petrol, and this latter event will often happen, although a fine strainer is always put for the petrol to go through before it enters the carburetter. This strainer is shown in Fig. 45, the central part being shown

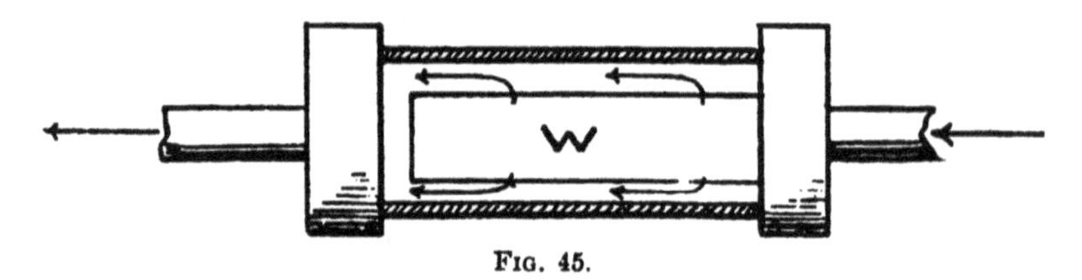

Fig. 45.

Wire Gauze Strainer for Petrol.

in section, and it consists of a gauze wire cylinder W attached to the end of the pipe through which the petrol enters, and this gauze cylinder is screwed into a larger metal cylinder, to which the other petrol pipe is attached. Any foreign substance, therefore, which might interfere with the proper working of the carburetter is caught in the gauze cylinder, and by periodically unscrewing the apparatus and taking it to pieces the accumulated dirt can be taken out and thrown away.

Reference was made in a previous chapter to the intense cold that was produced by the evaporation of petrol, in fact, the carburetter is a freezing machine. Even in hot summer weather, should there be any dampness in the air, the metal parts of the carburetter will become covered with a fine sort of hoar frost, which may ultimately interfere with its proper working. In order to rectify, this, a certain portion of the air drawn in is taken from a jacket round the hot pipe through which the exhaust passes, or in some cases, a part of the exhaust itself is arranged to go through a jacket in the case of the carburetter, so as to warm it up. The amount of artificial heating thus given to the carburetter depends on the state of the weather at the time, and it is arranged so as to be regulated at will By placing the hand on the pipe leading from the carburetter to the cylinder, its temperature can be felt, and if too cold it can be warmed by admitting more hot air, or exhaust gas, as the case may be, but as long as the petrol is being properly evaporated, the colder the mixture the greater power will the engine give, because each charge will contain more air and vapour.

When properly set these jet carburetters will work for hours without any adjustment being required, and they have the advantage of using every drop of petrol in the main tank up to the last drop. The petrol flows from the main tank to the carburetter either by gravity, or by air being pumped into the main tank, so as to force it up to the carburetter. In the latter case, of course, the main tank may be under the floor or platform.

This latter method of feeding the carburetter lends itself towards doing away with most of the risk of using petrol afloat, because the main supply can be fitted outside the boat altogether, such as a hollow keel, or in an external metal tube or tank, and its contents can be raised up to the carburetter by air being pumped in on top of the petrol in the main supply. The two pipes leading to the main supply, one for the air and the other for the petrol, must be carefully fitted and guarded against any external damage, which is always liable to happen to a vessel, and what has to be specially guarded against is that any overflow from the carburetter, arising from the valve not properly closing for instance, shall not be able to find its way down into the boat, but it must have a special pipe to lead to overboard otherwise the other precautions will have been taken in vain.

In Chapter X two methods of using petrol with a minimum of risk will be shown.

CHAPTER VII.

SUMMARY.

PROPELLERS—REVERSIBLE AND REVERSIBLE-BLADED—THEIR ADVANTAGES AND DISADVANTAGES—DAIMLER & HILLIER REVERSING GEAR—LOZIER—SEAL & VOSPER REVERSIBLE-BLADED PROPELLERS—PROPULSEUR AMOVIBLE.

We have now to consider the various patterns of propellers used by the several types of oil engines hitherto described; and the chief difference between those thus used and those used for steam, arises from the fact that the oil engine is not, strictly speaking, reversible. It is true, as was pointed out, that it is possible to reverse a two-cycle oil engine, but, as a matter of fact, it is too tricky a job to be able to trust to when coming alongside anywhere, such as a pier or yacht, so oil engines generally make use of a reversing propeller, that is the engine continues to run in the same direction as before, but the action of the propellor is reversed.

There are two ways of doing this, one being to cause the tail shaft and screw to turn round in the water in the opposite direction by means of reversing gear being fitted to the shaft between the engine and the tail shaft, and these are called *reversible propellers*. The other method is to twist round the blades of the screw on the central boss, so that, while still revolving in the same direction, their action on the water will be reversed, so as to give the boat sternway instead of headway. These are called *reversible-bladed propellers*.

The advantages of the reversible propeller are:—(1) The simplicity and small size, and consequent increased efficiency and strength of the screw and its boss; and (2) the greater facility of being able to get rid of weeds which have fouled the screw. We all know that at certain times of the year the weeds, both in fresh and salt water, are a positive curse to screw boats, and the quickest way of getting rid of them is by rapidly going ahead and astern with the screw. Now if only the angle of the blades of the screw is reversed, the weeds are not so easily thrown off, and they may have to be pulled off by a special cutting boat-hook made for the purpose, and this is a trying operation even in a shallow boat, whilst in a deep boat it is almost impossible to do, especially in a seaway or in the dark. (3) The saving of muscular exertion in starting the engine by hand, as only the engine has to be moved. This makes a great difference in an engine of any great power. The disadvantages of this system of reversing are that we have to add to the machinery a more or less heavy and bulky reversing gear, which, in many cases, is rather noisy, and another important point is that it is very difficult to prevent over loading the engine should we reverse the screw when the boat has a good deal of way on. It is also impossible to ease the pitch of the screw should the engine require it.

The advantages of the reversible-bladed propeller are:—(1) The absence of the above-mentioned heavy and bulky reversing gear, and the substitution of only a lever and a sliding collar or two; (2) the facility of being able to replace a damaged blade in a few minutes (when the boat is ashore, dry; (3) the facility of being able to make the pitch of the screw a little easier when going astern than it is for going ahead; (4) the facility of being able to ease the pitch of the screw when going ahead, should the engines require this, and, in some makes of screw, being able to set the blades fore and aft for sailing.

The disadvantages may be said to be that they require a larger boss and more clumsy shape of propeller, which is not so well able to stand a blow or wrench, and, from its shape, is often liable to pick up weeds.

In trying to strike a balance between all these advantages and disadvantages, we may say that for boats with deeply submerged screws, and more likely to cruise in deep waters, the reversible propeller will probably be the best; whilst for shallow boats with easily get-at-able screws, and where the saving of weight is of great importance, the reversible-bladed propeller will probably serve best.

Owners of boats with reversible propellers have often complained of their engines stopping when the reversible propeller was reversed, and the reason is that the pitch of the screw is so arranged that the engine when going properly is just able to drive it, whilst running at its proper revolutions. When the screw is reversed, the engine not only has the extra friction of the reversing gear to fight against, but also the headway of the boat, and this extra strain is often too much for it, especially if it does not happen to be doing its best; and all oil engines are like horses, they seem to have their days, one day they will run very freely, and perhaps the next day there will be a sensible falling off in the revolutions, although we do all that we can think of to rectify matters.

As was stated above, in the reversible bladed-propeller we can easily arrange for the pitch to be easier on the engine when reversed, and another very useful thing to have is the power to ease the pitch going ahead; for suppose, in a two-cylinder engine, that one cylinder will not work, or suppose that there is a certain amount of weed in the screw that we cannot get rid of, or even that we are screwing against a steep head sea and strong wind; then by easing the pitch, we can make sure that the engine will not stop, which is very likely to happen without such assistance.

The combination of a two or three-cylinder engine and a reversible-bladed propeller, in which the pitch can be varied at will, constitutes about the most reliable form of oil engine for marine work, as it would take an

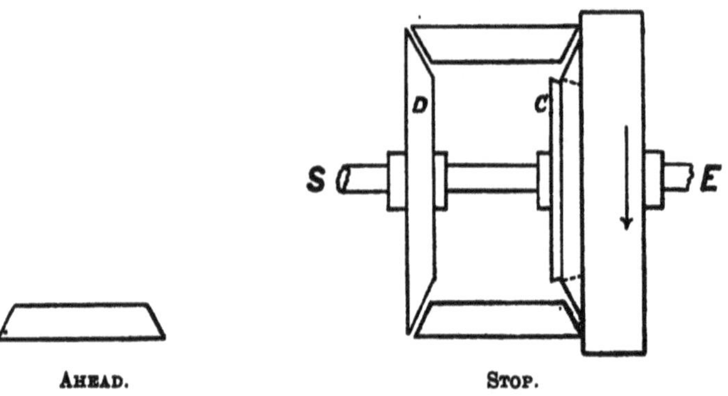

AHEAD. STOP. ASTERN.

FIG. 46.

DAIMLER REVERSING GEAR.

extraordinary combination of accidents to throw a launch thus fitted entirely out of gear.

The Daimler Co. introduced a very reliable form of reversible propeller which they have always stuck to, and it is shown in Fig. 46, where E is the engine shaft terminating in the flywheel, and S is the tail shaft carrying the propeller. The flywheel has a hollow cone seating turned in its after side, and the friction cone C can be pressed against this by a hand lever (not shown). This cone is fixed on to the tail shaft by a sleeve, so that although it is free to slide fore and aft for an inch or two, when it is pushed against the seating in the flywheel, the latter drives it and the tail shaft and screw for going ahead. By a slight motion of the hand lever, the cone C is

g. 50 shows another means of getting these results by means of a gear, and is used in the Vosper engines.

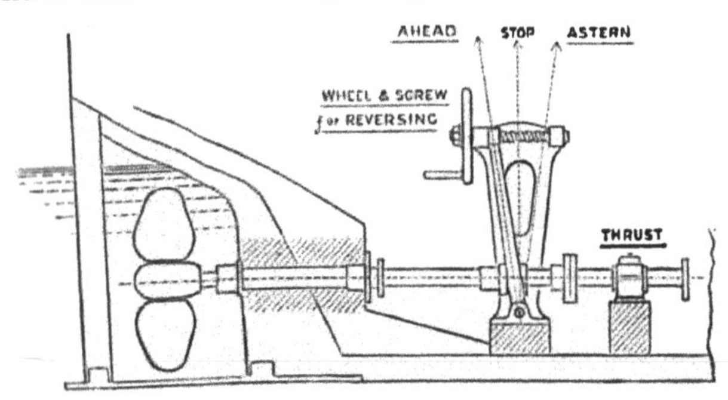

FIG. 50.

VOSPER REVERSING GEAR.

g. 51 shows a very ingenious form of propeller, which comes from , and which possesses a striking variety from all other forms hitherto here. The chief reason for describing it here is that it alone renders ible to apply an oil engine as motive power to racing boats, and, if it be brought out in a cheap and practical form, there might be a

FIG. 51.
PROPULSEUR AMOVIBLE.

erable demand for it for this purpose. The whole of our coasts are retty well provided with racing boats of fixed type, and they vary he humble "Redwing" in the Solent up to the powerful boats of t. Now the essence of these boats being of a fixed type is that they

withdrawn from the flywheel, and the boat stops, although the engine continues to run. In order to go astern, a further motion of the hand lever now brings in the two side bevel wheels (which are coated with rubber), and they are pressed against the after bevel of the flywheel, and the forward bevel of the wheel D. By this means the screw is gradually reversed, and, should the engine show any signs of being overloaded, the lever is slacked a little so as to let the wheels slip a little. This gear works very well and quietly, but care has to be taken that there is not any oil or grease on the bevel faces, or they would slip too easily.

FIG. 47.

HILLIER REVERSING GEAR. (OLD PATTERN).

Fig. 47 shows a somewhat similar reversing gear, as formerly used in the Hillier engines, and which may still be met with. It will be seen that the tail shaft is connected to the engine shaft by a sliding clutch. On moving the hand lever, this clutch is disengaged, and then the bevel gear wheel below is raised up so as to mesh with the two other bevel gear wheels, and thus the motion of the tail shaft and screw is reversed. It requires a little knack to strike in the gear neatly, and, although when going astern, the noise of the gear is evident, in going ahead, of course, there is no noise, as the lower gear wheel is then free. The later patterns of Hillier engines are now fitted with a very neat and quiet working reversing clutch gear, in which two small friction clutches are used with a gear wheel running in oil. The use of the friction clutches permits the head or stern gear to be struck in quietly and gently, and the overloading of the engine is avoided by this means.

Fig. 48 shows the action of the reversible-bladed propeller as used by the Lozier engine, and which is similar to that used by the Popular, and other two-cycle engines made in America. It will be seen that the tail shaft is hollow, and by means of the lever the outer portion is made to slide over the inner in a fore and aft direction, and by this means the blades of the screw are twisted round on their axes so as to give either headway or sternway to the boat, and when they are twisted so as to lie athwartships they only churn the water and do not move the boat.

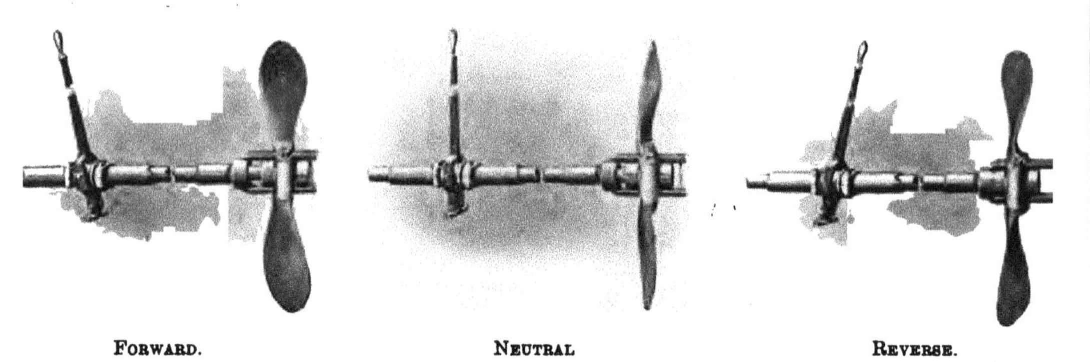

FORWARD. NEUTRAL REVERSE.

FIG. 48.

REVERSIBLE BLADED PROPELLER.

As the whole of the torsional strain is on the inner part, it will be seen that the outer part (and consequently the stern tube) has to be larger than is the case with the reversible propellers; moreover, there have to be two packing glands, one to keep the water from entering the outer portion and the stern tube, and the other to keep the water from entering between the inner and outer parts. Should the shaft get bent from any cause, the screw could not be altered, as the inner part would not slide in the outer.

In order to get over the foregoing disadvan
solid tail shaft connected to the engine shaft by a
moved forward or aft a few inches. The outsid
screw is prevented from sliding forward or aft by
part of the stern tube projecting into the water
the boat, round which, however, it is quite free to
solid tail shaft in or out, therefore, the blades are
the case of Fig. 48. Fig. 49 shows such a prop
engine.

It is very important that the sliding part c
firmly locked in the position given to it because it

FIG. 49

SEAL ENGINE, WITH REVERSIBLE BLADED

when running before a following sea the boat ma
and the tendency of this is to suddenly reverse
Another important point (especially in the larger si
blades should be gradually twisted from one po
when the engine is running, the effect of twisting
astern is very like that of gybing a mainsail, *i.e.*, a
blades get over the middle position they take char
lever almost out of the hand of the operator, thi
when the boss of the propeller moves fore and aft
Seal engine this is avoided by using an eccentric
twists the blades round without shock, and withou
By inserting a packing piece at the end of the
screw can be altered so as to suit the requirements

screw

Fran
know
it pos
could

consi
now
from
Belfa

should be more or less good sea boats, and suitable for occasional cruising, which the present racing boats are not, and for this purpose, as well as the being able to get to a regatta and back again in light winds, the addition of a motor would be of great advantage. Now they could easily be provided with the motor shown above, and, on arriving at the place of racing the motor could be unshipped bodily, and either put into a punt towed astern or into the boat of a friend until the return home in the evening, when it would be just as easily shipped again for the journey.

Another use for a separate motor such as this, is that any owner of one can go down let us say either to the Norfolk Broads, or to any part of the coast for a holiday, and hire a local boat for cruising or fishing, and without altering in any way the boat or her gear, can turn her into an auxiliary during the time of hire.

As will be seen, it consists of an oil engine lashed down on the deck aft, and which drives the screw by means of two bevel gears, just as the cranks drive the back wheel of a chainless bicycle. A peculiar feature is that the screw can be brought round so as to drive in any direction, so that it will steer the boat or even drive it astern, according to the direction of the screw, which is given by the controlling wheel shown at the side of of the engine; in fact a boat thus fitted would be just as handy as one fitted with twin screws so that she could be spun round without head or stern way at will.

Another very curious result is seen when the controlling wheel is let go. The screw now not only revolves on its axial shaft, but, whilst revolving thus, it keeps on turning round all the points of the compass. As the boat therefore does not know in which direction to go, it remains stationary, until the now revolving controlling wheel is grasped by the hand, when she starts off in the direction in which the screw is driving.

The engines and screws complete are made of the following sizes, *i.e.*, $1\frac{3}{4}$, $3\frac{1}{2}$, $4\frac{1}{2}$, 6, $8\frac{1}{2}$, h.-p., and the weights of the complete apparatus are 132 lbs. for the smallest and 308 lbs. for the largest. They can be obtained through the Paris Yachting Agency, of 4, Rue Meyerbeer, Paris.

It was stated previously that some makes of reversible-bladed propellors provided for the blades being able to be set to lie fore and aft so as not to impede a boat when under sail. This is a very important matter for an auxiliary, but in most of their screws the fore and aft position is only got by sacrificing the thwartship position. Mr. James Stephens of Stonehouse, Gloucestershire, makes however a propeller with all four positions of blades.

When the propellor used cannot set its blades amidships, a small clutch has to be fitted between the engine and the tail shaft. Very neat friction clutches can now be obtained, but still, they add to the cost and take up room, besides adding a little more weight. The author has in his boat a very neat and compact friction clutch provided by Messrs Tolch, of Fulham, and Messrs Vosper, of Portsmouth, also have a very compact and powerful one.

A friction clutch fitted in between the engine and the tail shaft will often save the screw being damaged when it fouls an obstacle, such as a rope or small piece of wreckage, if it be adjusted so as to only just convey the full power of the engine; any extra strain will cause it to give a little and thus ease the shock. Owing to the weight of the flywheel of an oil engine, a tremendous strain is brought on to the tail shaft and screw when an obstacle is suddenly fouled by the latter.

In a large engine, it is a great thing to be able to disconnect the screw for starting, as then only the resistance of the engine is felt when turning by hand.

As the correct pitch of a screw driven by an engine of given power depends on the speed of the *boat* as well as that of the *engine*, it is difficult to know what pitch to give the screw when fitting an oil engine in a sailing boat, which, of course, will not go at anything like the speed of the small launches for which the screw supplied with the engines is designed. The best plan in this case is to borrow a screw with adjustable blades, and by trying several experiments the best pitch for the particular boat and engine can thus be found out, and then a properly designed screw can be obtained from any skilled firm of engineers, which will absorb the same energy from the engine, and give slightly better results as to speed than were obtained with the adjustable screw.

A ball-bearing thrust is a very useful means of getting increased efficiency from an engine, but for salt water, the steel balls want very careful looking after, and it is better to keep them smothered in stiff grease, than to use oil and they should always be enclosed in a cage, so that they cannot get loose and tumble down in the boat

CHAPTER VIII.

SUMMARY.

IGNITION—VARIOUS SYSTEMS—WYDTS CATALYTIC IGNITER—DRY CELLS—ACCUMULATORS—ELEMENTARY PRINCIPLES OF THE ELECTRIC CURRENT—VOLMETERS—AMMETERS—LOW TENSION BREAK-SPARK—SPARKING COILS—SYSTEM OF WIRING—MAGNETO SPARKERS—SIMMS BOSCH MAGNETO IGNITION—HIGH TENSION JUMP-SPARK—COIL & TREMBLER-SYSTEM OF WIRING—SWITCHES-IGNITION PLUGS—GENERAL HINTS.

We now have to deal with the different systems of igniting the mixture in the cylinders of the engines we have previously described, and the methods generally used may be divided into three classes, *i.e.* :—

Ignition by hot tube.
Ignition by stored up heat.
Ignition by the electric spark.

The hot tube ignition has already been described in these articles, so that we need not enter into it any more here, but will commence with the second class. In this form of ignition a special substance, such as a piece of porcelain or platinum, is placed in communication with the combustion chamber, so that after it has been heated up to a certain temperature by some means or other it is hot enough to fire the mixture when it is forced into contact with its glowing particles by the compression stroke, and the repeated explosions thus produced suffice to keep up the necessary temperature, so that when once started the engine goes on working continuously.

The best known engine using this system is, perhaps, the Hornsby Akroyd, which, using paraffin oil, is now extensively used for country house and farm work. In this engine a cast iron chamber is bolted on to the back of the combustion chamber, with which it communicates. On first starting, this chamber, is heated to a dull red heat by a paraffin lamp worked by a hand blower. The engine is then turned round by hand, and the explosions take place as stated above, so that the engine will go on working the whole day, or as long as it is supplied with oil. The paraffin launch 'Abiel Abbot Low,' which recently came across the Atlantic, had this form of ignition for its 10 h.p. two-cylinder engine, but many engines over here, after being started, get their subsequent explosions by means of an *igniter* so fitted that it gets its heat from the first five minutes or so of working, and when thoroughly heated, the lamp which heats the tube can be extinguished, and the engine goes on working. In some engines this igniter is made of a piece of porcelain, in others a sort of nest of iron wire is used, and several forms of platinum, both in the wire form and of the spongy nature, have been used for the purpose. Unless the engine runs at a considerable speed it is difficult to keep up the necessary heat in the igniter, and if the governor acts by cutting out explosions, sometimes the igniter will cool down and fail to act. The chief objection, however, is that if the engine is stopped for any cause, then, in order to start again, the whole process of lighting up the lamp has to be gone through again in order to reheat the igniter.

In the case of a petrol engine with electric ignition, of course there would not be much trouble in starting the engine again after a stop by means of the electric spark; but as the trouble and expense of the electric installation have been incurred, owners generally prefer to use the electric ignition continuously.

There is, however, so much trouble and worry attached to even the best form of electric ignition, that the author fully believes the ignition of the future will be by some form of igniter, probably heated at first by electric means. This has been the dream of experimenters for many years, and it may be worth while to explain the most recent invention in this direction.

Fig. 52 shows Wydt's Electro-Catalytic Igniter all ready to be screwed into the place for the usual sparking plug. The rod D slides in and out within the body K, and makes a gastight fit. At its inner end is a small coil of an alloy of platinum, osmium and rhodium, the two latter elements being rare metals much akin to platinum, and, on first starting, a current of electricity is sent through this coil, which renders it incandescent. The engine is now started, and the succesive explosions keep the coil incandescent without the electric current. The timing of the ignition is obtained by sliding the rod (with its coil) in or out within the body, and to stop the engine suddenly the rod is so far withdrawn as to uncover the small passage E communicating with the outer air. The first suction stroke of the engine then draws in a charge of cool air on to the coil and cools it down at once. Of course, if the engine be stopped it is no trouble to heat up almost instantaneously the osmium coil again. Experience alone will show if the coil will deteriorate with use and if the sliding joint will let the gas leak, but if, not, then this form of igniter should prove very useful, as it only uses the electric current obtainable from one dry cell for about 15 secs. for starting, whilst there are no *contact points* to corrode and burn away, which (as will be shown under the description of electricity) are often a source of stoppages and annoyance with electric ignition pure and simple.

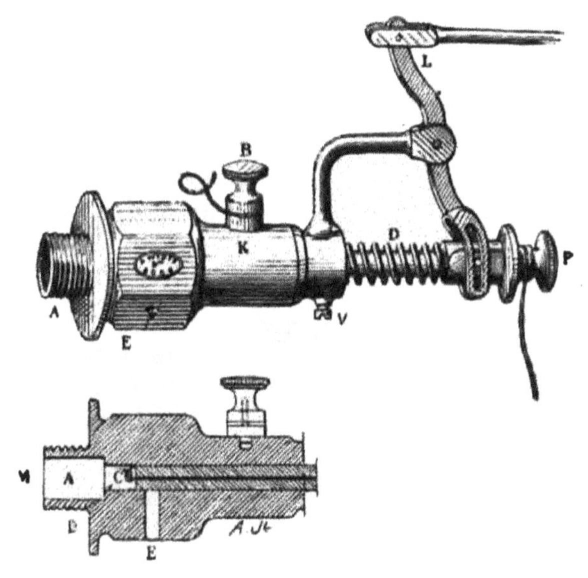

Fig. 52.

Catalytic Igniter.

We will now proceed to the more serious part of describing the electric ignition, and, seeing that pretty well all petrol engines now use this form of ignition, it is only stating a fact to say that all users of these motors ought to understand something of the nature of, and general principles involved in the use of the electric current. If all that they know about it is that somehow or another it makes the spark in the cylinder, then sooner or later they will find themselves with an engine that will not work, and will be quite helpless as to getting it right again, although, perhaps, with the necessary knowledge, it would only be the work of a minute to discover the cause and rectify it.

Electric ignition may be divided into two forms as now used, *i.e.* :

The low tension break-spark.
The high tension jump-spark.

Before attempting to go into the difference between these two, we must first of all study electricity itself a little.

We cannot, of course, here do more than just describe the rules governing the use of electricity in oil engines, and some of its peculiarities which render it so capricious in use.

To begin with, we generally buy our electricity either in the form of a dry battery, or in an accumulator. The dry battery is more or less similar to the Leclanché which rings our household bells, only the liquid is replaced by a jelly-like substance, and one of four cells is shown in Fig. 53 and when first bought it is ready to give out a current for so many hours, and when it

Fig. 53. Dry Cells.

Fig. 54. Accumulator.

is exhausted or run down it is practically useless and has to be thrown away. The accumulator of two cells shown in Fig. 54, on the contrary, when run down can be charged for us again, as many times as we wish, at any electric-charging station, and the usual charge for this is one shilling. We can also re-charge the accumulator ourselves by means of suitable batteries.

Before going into further detail as to these two sources of electricity, we shall have to consider the way in which electricity itself is measured, and for this purpose it will probably assist if we try to compare electricity with some well-known substance such as water.

If we have a tank containing water just under five feet deep, and we fit a pipe P, Fig. 55, to the bottom of it, we shall find that there is a pressure in this pipe of 2 lbs. per sq. inch, and if we measure the quantity of water running out of this pipe in a given time, we shall find, if the pipe is of a certain size, that, say, one cubic feet of water runs out in an hour.

Now, in dealing with water, as above, we make use of two units of measurement, *i.e.*, the pound as a unit of pressure, and the cubic foot as a measure of quantity. In dealing with electricity we require two similar measures, and these are the *volt*, or unit of pressure, and the *ampère*, or unit of quantity. Of course there is no connection between the pound and the volt, and the cubic foot and the ampère, but they are similar measures used for the two different substances.

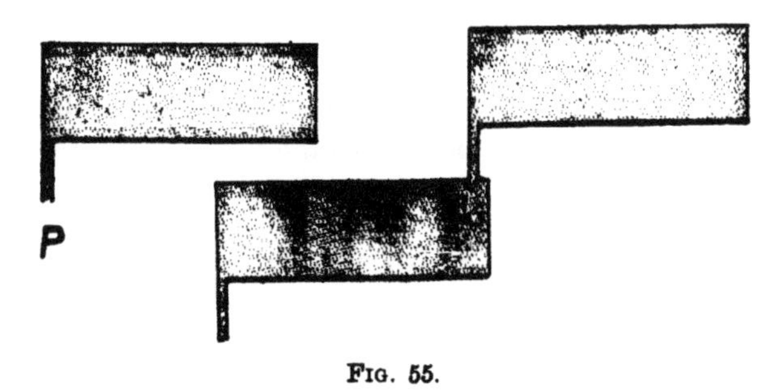

Fig. 55.

Pressure of Water and of the Electric Current.

The pressure of the accumulator, when first charged, is about 2·2 volts per cell, and when first used, this pressure soon falls down to two volts per cell. It remains at about this pressure for the rest of the time in use, until, just before getting exhausted, it suddenly drops down to about 1·8 volts per cell. It should never be used after this drop, but should be re-charged again as soon as possible.

The pressure of dry cells varies from about 1·2 to 1·6 volts per cell, and they may be used as long as they will give the necessary spark, as they have to be thrown away when used up. It is worthy of note that they can be given a short, fresh life when exhausted by injecting into each cell a strong solution of sal ammoniac (such as is used in Leclanché cells). The knowledge of this may perhaps sometimes enable one to bring home a boat when the battery has run down.

The amount of electricity in a battery or accumulator is expressed in ampère-hours. Thus, say, we are told that a battery is good for 30 ampère-hours; this means that it will give a current (at a useful voltage) of one ampère for 30 hours. If we ask it to give a current of two ampères it will do so for 15 hours, and so on.

This does not mean that we could only run the engine for 15 hours (supposing the coil required the two ampères to work it properly), because we must always remember that the current only passes for a very short portion of the time; thus if the current only passed during a quarter of each revolution of the flywheel, then the battery should run the engine for 120 hours if 1 ampère were enough.

With each battery or accumulator there are two terminal wires, one marked + known as the positive, and the other marked — known as the negative; and the current is always trying to run out of the + wire into the — wire, so that if we connect them by any conductor, such as wire, the current will flow through it. The + wire, therefore, may be said to represent the pipe P above. If we disconnect the wire, the current cannot flow, and it is the same as if we had shut off the pipe P by a stop cock.

Referring again for the time to our water tank. If we connect the outlet pipe of one tank on to the top of another similar tank, as shown on the right in Fig. 55, we should find that the pressure in the lower pipe P would be just double what it was before (with the single tank). Similarly, if we were to thus connect up three tanks, we should find that the pressure was three times the original amount. Now this is exactly what happens, as regards the voltage, if we connect up two or more electric cells, by joining the — terminal of one to the + wire of the next and so on. This method of joining is known as *connecting in series*. The accumulators, as we buy them usually consist of two cells in series, which, therefore, when charged, give us a total current of 4 volts pressure, and a little over.

The dry cells only give 1·3 to 1·6 volts per cell as a rule when new, so that we have to use three or more connected up as above, or *in series*, and we get from the three about 4·3 volts.

If we were to let the pipe P drain out the water from the tank we should find that both the pressure and the quantity would gradually get less and less, and this is exactly similar to what does happen in the case of the accumulator or battery. We can recharge the former, and it is as if we had pumped up the tank full of water again. It would not hurt the tank of water if we ran it out dry, but this must *never* be done to an accumulator, or it will be ruined. They should never be run down below 1·8 volts per cell, and as soon as they only give this current, or 3·6 volts together, they should be recharged. In order to know what is going on in an electric installation we require to be able to measure the voltage and ampèrage, and this is done by the voltmeter and ampèremeter or ammeter (Fig 56). They are simply connected up to any two terminals, between which the current is trying to pass, and the reading is given on the dial as in the case of a pressure gauge.

Note particularly that the ampèremeter should *never* be connected up to the terminals of an accumulator or battery *direct*, nor should the terminals ever be allowed to touch each other, as if this is done, it will damage the battery and probably ruin the accumulator. In order to get the ampèr-

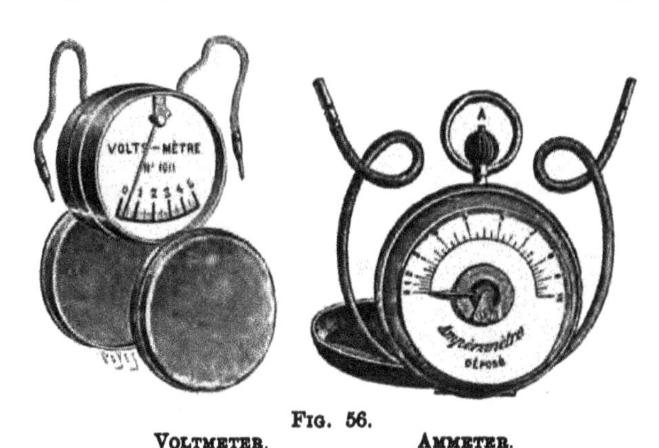

Fig. 56.
Voltmeter. Ammeter.

age we must always see that the current is going through some extra resistance, such as a sparking coil, for instance, before we apply the ampèremeter. No harm is done if we connect up the voltmeter to the terminals direct.

Supposing, now, that we have our source of current all ready, the thing to be considered is how to cause it to make the spark in the cylinder when required. We have just explained that the terminals of a battery or accumulator should never be brought into contact without some resistance between them, but if this were done, as an experiment, we should find that there was a tiny spark shown at the moment of *touching*, and a larger and more brilliant one at the moment of *breaking* contact. If we now, however, interpose between the terminals a spark coil, which consists of a great many coils of wire wound round some soft iron rods, we shall find that the spark visible on making and breaking contact is far larger and more brilliant. And if we use a series of batteries or accumulators which will give a current of about 6 or 8 volts, and send it intermittently through a proper spark coil, we shall have (at the moment of breaking contact) a spark large enough to fire the mixture in the cylinder. This is the so-called *low tension break spark system* and it is that generally used in the American two-cycle engines and those of similar nature, the break of contact takes place in the cylinder. It is more wasteful of electricity than is the high tension jump spark ; and, moreover, is not so suitable as the latter for very fast-running engines. But its greater simplicity and reliability have rendered it very popular where suitable.

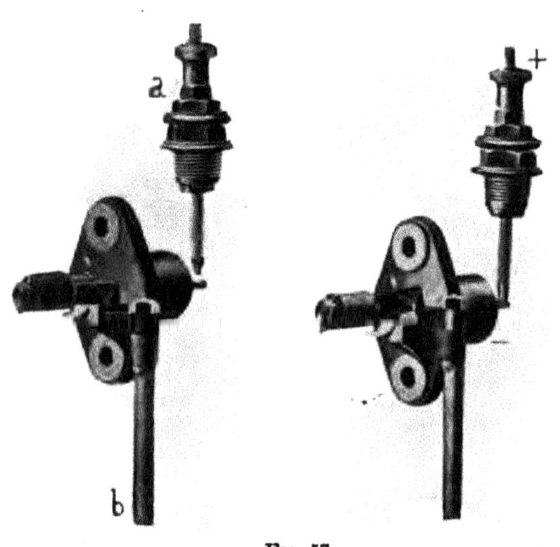

FIG 57.
LOW TENSION. SPARKING POINT.

We must now describe the mechanical parts which make and break contact at the proper moment.

The actual spark is produced between two contact pieces projecting into the cylinder; they are made of steel or nickel, and tipped with platinum just where the spark appears. One of the pieces is surrounded with insulating material, such as porcelain or mica, and is fastened in the head of the cylinder generally ; it is known as the insulated plug. Fig. 57 shows these pieces as fitted in the Truscott Motor. The other piece is called the rocking arm, and is fitted so that as it rocks in its bearing its inner end alternately touches and leaves the end of the fixed piece. This rocking arm has to move freely in its collar, and yet must be a gas-tight fit.

Now supposing these two pieces connected up to the source of electricity, so that the current is trying to get from the insulated plug to the rocking arm, as long as they are separated ever so little the current cannot get across the interval. Supposing the piston to be rising, and that it first of all causes the rocking arm to touch the insulated plug, then the current begins to flow, but the small spark thus made is not large or hot enough to fire the mixture. Just before the piston reaches the top stroke, however, it *suddenly* pulls away the rocking arm, and a large fat spark jumps across the separated platinum points, and this fires the mixture. We say suddenly, because the quicker the points are separated the hotter is the ensuing spark.

The mechanism for producing this motion of the rocking arm varies according to the make of engine. There is nearly always a vertical rod connected to an eccentric or cam on the main shaft of the engine, so that at every revolution this rod goes up and down, and a cross piece at its upper end engages with the tripping piece of the rocking arm, lifts it up, and then lets it go again suddenly. (b), Fig. 57, is the upper end of this rod. In some cases this rod is pivoted about its middle, so that its upper end describes a sort of oval path. In others there is an arrangement like the trigger of a gun which comes into contact with a fixed screw. This latter then pulls the trigger at the proper moment. Moreover, by raising or lowering very slightly the end of this screw we can cause the trigger to be pulled either early or late so as to time the ignition within certain limits.

By turning round slowly the fly-wheel of any engine fitted with this form of ignition we shall easily see the successive movements, *i.e.*, first of all the tripping piece of the rocking arm will be moved so as to bring into contact the two sparking points within the cylinder, and its sudden release will indicate the moment when firing takes place.

The wire connected to the + terminal of the accumulator is connected to the piece a, and the other, or – wire, is connected to any part of the engine, and the electric current finds its way back from the rocking arm, through the whole mass of metal composing the engine, bed-plate, etc., to the negative wire. This latter wire is then said to be connected *to earth* (because in land telegraphy the earth is thus used to convey the return current).

A switch is generally provided on the cylinder head, so that the current can be completely cut off, and thus any leakage, when the engine is not running, is guarded against. This switch should always be open whenever the engine is stopped, if only for a short time. Another switch should also be fitted close to the battery, which should also be left open, when the boat is put away on the moorings.

Fig. 58 shows the sparking coil. It consists, as was stated before, of several turns of carefully insulated wire, coiled round a mass of soft iron

Fig. 58.
Sparking Coil.

wires, and a vulcanite case protects it from the effect of the wet. It should, however, be always placed in a sheltered part of the boat.

Fig. 59 shows the method of connecting up the battery, coil and engine, where B shows the battery, C the sparking coil, S the engine

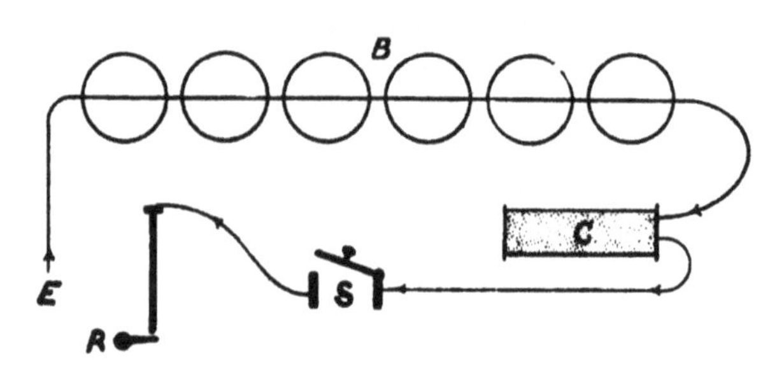

Fig. 59.
Wiring for Low Tension Break-Spark

switch, R the rocking arm just separated from the insulated plog, and E the other terminal of the battery connected to the metal of the engine, or earthed, as it is called, so that the current can find its way from the rocking arm back to the battery. This system of wiring should always be carefully adhered to.

It is very easy to make a mistake in doing this, and thus let all the electricity escape, damage the battery, or probably ruin the accumulator. Getting the electric current to go just where we wish it to, and nowhere else, is something like driving a pig to market, we make the course of the current as clear and easy as we can, but it seems to seek out every other possible way of going, as if on purpose sometimes. Should any wet get on the covering of the insulated wires, the current will run along outside along the moisture instead of to the sparking points. Should the insulating material of the plug in the head of the cylinder get cracked (which is very likely to happen, either from a blow or from the effect of the heat of explosion) a deposit of soot in this crack will give the electric current an excuse to escape thereby, and thus shirk its proper work.

After a time the platinum sparking points inside the cylinder are very likely to get burnt and corroded, and then the current refuses to jump at all, and so misfires result. It may be stated as a general rule that, in nine cases out of ten breakdowns, the source of trouble will be electric.

When we first get the engine all right, and when everything is in good working order and new, we should always go through the following tests, and make careful notes in a small pocket book of the voltage at the terminals of the battery, the voltage and ampèrage of the current passing through the coil and sparking points and engine from the battery. By having these figures carefully noted we can always at any time check them afloat, and when there happens to be a breakdown, or the engine is not running properly, we shall often find great assistance thus, in locating the source of trouble.

A mark should be made on the flywheel (if the makers have not already put one) to show exactly when the sparking points begin to touch each other, and when they are separated. At any time afterwards then if we have to file away the points, or put new ones in, we can ensure the proper adjustment of these two most important items.

Supposing we are afloat, and the engine either stops or refuses to start. The way to investigate the cause would be (1) Try the voltmeter across the terminals of the battery or accumulator ; this should show about the proper voltage as noted in our pocket book if this is right (2) Disconnect the – wire from the mass of the engine and strike it along the screw of the insulated plug, seeing that the switch on the top of the cylinder is closed. A series of good sparks should be visible. If these do not show it will point out a defect or breakage in the wires leading from the battery through the coil to the engine, or a defect in the coil itself. To test for the latter, arrange a wire so as to send a current through the coil from the battery, and test with the voltmeter and ampèremeter, the current thus passing should not be less in volts and ampères than that noted down as the correct amount in our note book. If it is much less, then the coil will be defective, and the boat had better be got home in the best way without the engine, as repairing a coil is a job only for the qualified electrician.

If, however, this current is all right, the defect must be looked for somewhere in the wires between the engine and the coil, as it cannot be anywhere else, and inspection will probably reveal either a broken wire, or perhaps the two wires have rubbed together and worn through the insulating cover, and thus the current will escape the short way home. Having now rectified this, we find that we get the good sparks by striking the wire as in (2).

Having thus proved that the current is entering the insulated plug all right, the thing to seek for is, perhaps, a crack in the insulating material, which permits the current to make a short cut, instead of passing between the sparking points. To test for this (3) see that the switch on the top of the cylinder is closed, and that the engine is in such a position that the cross piece at the top of the rod has not yet touched the tripping piece, then try if any current can be got from any part of the mass of the engine. Even if no sparks can be got, try and see with the voltmeter if any current is passing, and if such is the case, the insulation of the sparking plug is either broken down, or a coating of soot has provided a passage for the current. No current at all should be able to pass from the insulated plug to the mass until the rocking arm has been moved, so as to make contact within the cylinder: but if such be the case, the sparking plug must be removed from the cylinder, and all soot carefully removed. Should the insulating material show any crack, a new plug must be inserted (spare ones and parts should always be carried afloat). The defective plug can be repacked and insulated at leisure.

The last thing (4) is to turn the engine round until the rocking arm is fully raised up, and just on the point of slipping off the cross piece. See that the switch on the top of the cylinder is closed, and try striking the wire on the outer end of the rocking arm, when a good spark should be visible. If no spark is shown, then the platinum points will probably have got burnt and corroded, and we must take them both out from the cylinder and clean them up with a very fine watchmaker's file. On putting them back again into place, and causing them to touch each other by turning the engine, we shall get the proper spark outside, and we may then connect up properly and feel certain that the engine will now work as well as ever, unless the fault lies in some other part of the engine, beyond the electrical ignition apparatus.

NOTE.—After having filed the platinum points, see that the make and break of contact take place at the proper time, as per the marks on the flywheel, and according to the instructions which are sent out with each engine.

The rocking arm must be a nice easy fit within its bearing, so that it will always act quickly and properly, and not hang up. It should occasionally be lubricated with a little paraffin oil. If it be pressed up by hand so as to make contact with the insulated plug, a slight motion in and out will rub the platinum points together, and often clean off any slight deposit of soot, which might interfere with the proper ignition. The points should be separated about $\frac{1}{16}$ in. when at the furthest distance apart.

It has been explained that the two usual sources of electricity, *i.e.*, the dry battery and the accumulator only lasted for a certain time, when they either had to be renewed or recharged. In order to avoid this trouble, makers have always tried to get the necessary supply of electricity from the engine itself by means of a *magneto sparker*. We cannot here go into the principle and construction of this instrument, but will only say that it consists of a revolving part, which is a soft iron armature covered with insulated wire, revolving more or less within the two poles of a set of permanent magnets of the usual horseshoe form. The actual ends of the magnets are extended by pole pieces of iron, so as to embrace the revolving armature.

One of the greatest discoveries of Faraday was the fact that such an armature, when caused to revolve by mechanical means, caused a current of electricity to flow in the wire surrounding it; and during each revolution there were four of such currents, *i.e*, two in one direction, and two in the opposite direction. As we want the current to flow only in one direction, or be *continuous*, there is an arrangement fitted at one side of the apparatus known as the *commutator*, which reverses every other current, so that we have a rapid succession of currents always in one direction, all ready to be conveyed to the sparking points by wires, just as in the case of the battery.

The magneto may be driven by means of a belt from the engine, or it may be fitted with a small friction wheel, which is pressed against the circumference of the flywheel, and thus a great number of revolutions are obtained. Of course, a certain amount of horse-power is absorbed in

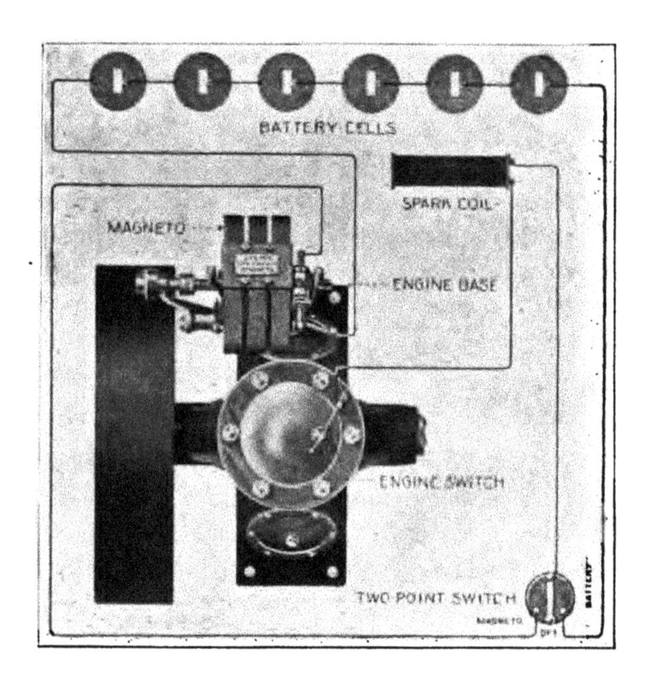

Fig. 60.

Wiring for Magneto and Battery.

driving the magneto, but the advantages of a constant source of electricity are considered to outweigh this very small loss of power.

Fig 60 shows a magneto driven by means of a friction wheel, and it will be seen that the armature revolves between the poles of two sets of magnets placed opposite each other. In some engines, magnets are affixed to the flywheel; and, as it revolves, the ends of the magnets are rapidly drawn across the armature, which is fixed; but the principle is the same. Fig. 61 shows a magneto sold by the Mitcham Co. The amount of current given out by these magnetos increase rapidly as the revolutions increase; and, although this is a good thing from one point of view, *i.e.*, that as the engine runs faster, there is less time for the electric current to get to its maximum and make its spark, so that, an increase of current up to a certain point is thus useful, yet a certain difficulty is introduced by this fact, in the starting of engines by hand (except in the smaller sizes), because, when turning the engine by hand, we can only do so slowly; and if the magneto is arranged

Fig. 61.
Magneto Sparker

to give the necessary spark at this low speed, then, when the engine is running at top speed, we shall get too much current, which would tend to overheat and damage the wiring of the armature. Owing to this, the magnetos are naturally arranged to give the proper spark when the engine is running at full speed, so when we try to start it by hand we cannot get the necessary spark, and the usual course is to use batteries for starting; and then, when the engine is running properly, a switch cuts off the battery current and cuts in that from the magneto. The arrangement of this will be understood by a further reference to Fig. 60. By only using the current from our batteries thus, of course, they will last a long time.

An ingenious means of doing away with the difficulty about starting with the magneto is afforded by an American invention, known as the Autosparker, shown in Fig. 62. It is arranged so that it will give the necessary spark when turning the engine by hand slowly, and the dangerous increase of current, when the engine is running at full speed, is prevented by means of a governor, shown at the end opposite the flywheel. As soon as the speed is getting too much, the governor slightly rocks the magneto on a central pivot, and thus allows the friction wheel to slip a little, until the proper speed is reached again. With this sparker, therefore, no batteries are required.

Fig. 62.
Autosparker.

All friction-driven magnetos should be mounted so to rock on a central pivot, like the Autosparker, so as to allow for any unevenness in the running of the flywheel; for, however true it is turned, there is often a tendency for it to wobble a little when running fast, especially when the counterweight (to balance the crank) is put in the flywheel. Care must always be taken not to allow any oil to get on the friction wheel, or it will not drive properly. Magnetos should always be so covered as to be proof against wet and spray which will often get on them afloat.

The commutator will require cleaning from time to time where the collecting brushes for the current press against it, and this can be done by means of the finest glass-paper or emery cloth. The brushes should be set so that hardly ever do they spark; if there is much sparking it will be certain to soon wear out and burn away the commutator and brushes. Carbon brushes give better results in this respect than those made of metal.

One of the simplest and most compact systems of magneto ignition is that known as the Simms Bosch magneto, and it will be seen in the diagram of the Simms engine, Fig. 63.

FIG. 63.
SIMMS-BOSCH IGNITION.

It works on a different principle to the magnetos described above, because both the magnets and the armature are fixtures; but between them is mounted a thin, soft iron envelope, with parts of it cut away, and this envelope is caused to *oscillate* by the engine instead of being revolved. Its action is as follows: With the apertures in the envelope in a certain position between the poles of the magnets and the armature, the latter becomes saturated with magnetism, but as long as we leave everything quiescent there will be no current in the wire round the armature. If, however, we suddenly rock the iron envelope so as to alter the position of its apertures between the poles of the magnets and the armature, the magnetic lines of force are cut and intercepted by the envelope, and the armature *suddenly* loses its magnetism, and becoming magnetised in a reverse direction a strong current of momentary duration appears in the armature wire, which is quite enough to fire the mixture. There is a certain position of the envelope when the current is at its maximum, and the two sparking points must be separated in the cylinder at exactly this moment. A very ingenious timing arrangement is comprised in the mechanism, so that the break may be made to occur either early or late, but always just at the period of maximum current.

It takes only a very little power to rock the envelope, which is very light, and, of course, there are no troubles from belts or friction wheels slipping, no commutator to clean, and no coil required. The circuit for the current is absolutely closed, the only break ever being made in it being at the sparking points. This system has now been tried for some years, and as its use is now extending to motor cars, it may be said to have served its apprenticeship and proved a practical invention.

There is one fault that all magnetos suffer from, and that is the magnets require to be remagnetised from time to time. All vibration and heat is bad for magnets, and these are just the two things that we cannot entirely avoid with an oil engine. The magnets should be so mounted and protected as to reduce these two disturbing elements to a minimum.

Having now described the low tension break-spark, we will proceed to the other form, or high tension jump-spark, which is almost universally used in motor cars, and will, therefore, often be met with afloat. The chief difference between this system and the other lies in the fact that the two sparking points are fixed in the combustion chamber a certain distance apart (generally about $\frac{1}{32}$ in. to $\frac{1}{64}$ in.), and that they never move; the spark has to be made to jump across the small gap between the points, and in order to get it to do this, the current has to be of a very high tension, amounting to many thousands of volts! Now, although we require a current of this very high tension, we only require a very little quantity or ampèrage, so that this system actually uses less current than does the other one. A two-cell accumulator, or four dry cells, combined with a proper sparking coil, will give the spark, and a properly wound coil should not require more than about one ampere at four volts to work it. A very great deal depends on the coil, the way it is wound, and its insulation, and especially its condenser. There is no falser economy for an oil engine than a cheap or inefficient coil. They are wound with two separated coils of insulated wire, the inner one being of a larger size and shorter, and known as the *primary;* and the other, which is wound round outside the primary, being of a much greater length, and of a finer wire, known as the *secondary.* The source of electricity (battery, accumulator or magneto), is connected up to the primary, and the secondry wire is connected to the sparking plug, which is screwed into the combustion chamber, and carries the two sparking points.

Now every time that the primary current is made and broken, another current of many thousand volts tension appears in the secondary wire, and owing to its high tension it is able to jump across the gap in the firing plug. It is customary in some engines to make and break the primary current several times for each ignition, so that, instead of only one spark jumping across the gap, we get a stream of sparks, and this renders the ignition more certain; because, if the first spark does not ignite the mixture, one of the subsequent ones will do so, and some persons believe that the effect of the first sparks is to render the subsequent ignition more easy.

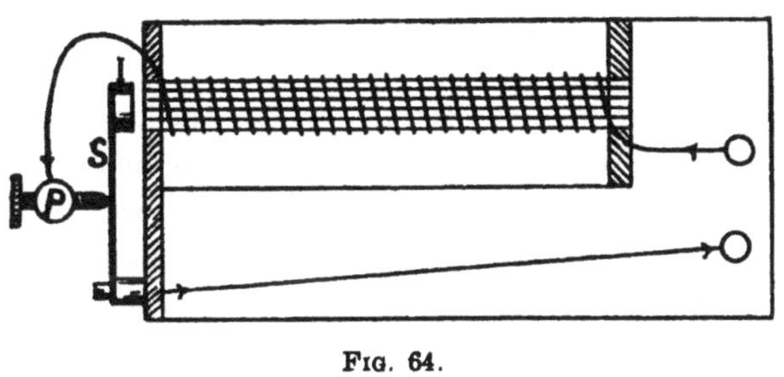

Fig. 64.

Coil, with Trembler.

The means of rapidly making and breaking the primary current is afforded by the *trembler,* which may be either mechanically, or electrically worked. The mechanical trembler, although much used for motor cars, is not often used for marine engines, so we will only say here that it is a sort of vibrating tongue of metal, which buzzes against a set screw, and thus causes an extremely rapid series of makes and breaks in the primary current. The electrical trembler is part of the coil, and its action depends on the fact that every time a current is sent through a coil of wire surrounding a bundle of soft iron wires, it makes the latter into a bundle of magnets. Reffering to Fig. 64, which shows a section of a primary circuit wire coiled round the soft iron wires of a coil (the secondary wire is not shown), S is a spring which presses against the set screw in the post P, and they form part of the circuit of the primary current. As soon as this current flows through the coil, the iron wires become magnets, and consequently attract the piece of iron I at the end of the spring, and pull away the spring from the set screw in the post P. This breaks the circuit, and directly this happens the iron wires no longer attract the iron on the spring so that it again presses against the set screw in post P, makes a fresh circuit, and the operation is repeated many times a second. The set screw in the post P regulates the tension of the spring, and consequently the number of vibrations per second, which latter can be calculated from the musical note that it gives out when working.

All the time that the spring is vibrating, a series of small sparks can be seen between the points of contact; and although they are tipped with platinum, yet, in time, even this metal becomes corroded, and requires cleaning. A condenser, consisting of a number of sheets of tin foil, properly insulated, is connected up to the terminals of the primary circuit, and diminishes the sparking a good deal, but still it, and its consequent effects, are always visible.

Every place in an electric circuit, where the current is broken, is a source of more or less trouble, owing to the burning and corrosion produced by the spark. Such places should be always kept free from oil, as if it gets burnt on the platinum it will stop the current from passing; periodically the platinums should be touched up with a very fine watchmaker's file.

As platinum is a very expensive metal, there is a great temptation to use cheaper substitutes, such as silver and German silver Whenever the spark shows a greenish tinge (instead of electric blue), it may be taken as showing that pure platinum is not used for the contacts.

Having now followed out the working of the primary current in the coil, we can understand that if the engine makes a contact so as to start the primary circuit at a proper moment in its revolution, the trembler will be caused to rapidly vibrate, and a stream of alternating sparks will jump backwards and forwards between the points of the firing plug. Fig. 67 shows how the engine makes the necessary contact. The cam below the spring I is connected to the engine shaft, for a two-cycle engine, or to the half-time shaft for a four-cycle engine, and at the proper moment, the spring is pressed up by the cam, so as to make the contact between it and the screw above: the primary current now starts, and the electrical trembler goes on until, by the revolution of the cam, the spring I breaks contact.

The spring and the contact screw are mounted on an insulating block (generally vulcanised fibre), which can be turned round by means of a link, through a certain angle. If we turn it to meet the cam, it will make the engine fire sooner, or advance the spark; if, on the contrary, we turn the block slightly in the direction the cam is revolving, this will retard the spark. By thus working the *timing gear*, we can alter the working of the engine a good deal.

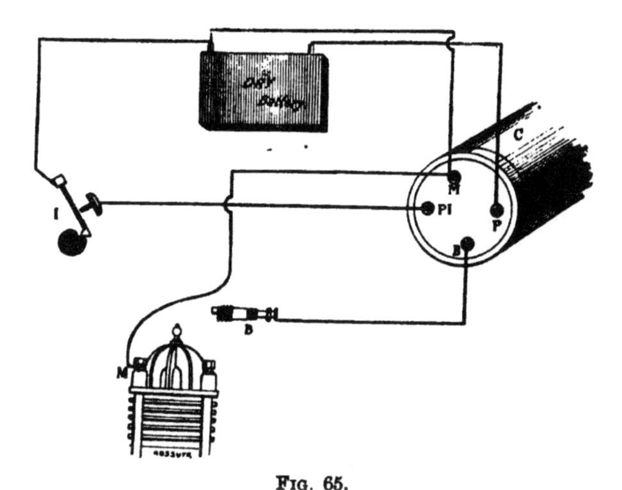

FIG. 65.
WIRING FOR HIGH TENSION JUMP-SPARK.

All coils are not made with the same terminals; some have only three, and in some a band round the outside represents one of the high tension terminals; so that the Fig. 65 must only be taken as a general explanation of the principle of the high tension system. Whenever a coil is bought anywhere, descriptions of the clearest sort must be asked for as to its proper connections; a single mistake in them may spoil the battery, or prevent the engine from firing, or, what is more puzzling to the amateur, may cause the engine to fire properly at one time, and not at another.

Owing to the high tension of the secondary current, the wire conveying it must be insulated with the very best material, or constant leakage will result; and the wire must be treated with great respect, as a shock from it is not at all pleasant. One need not fear it, however, as being actually dangerous. This high tension current is very difficult to keep within bounds, and, in wet weather, when everything in a boat becomes damp, it will often escape and run down the moisture outside of the wire, and thus shirk its work. When working an engine on a wet and dark night in a boat, it is astonishing to see the electric stream thus escaping, and it is often enough to give one a perceptible shock. Whenever wires have to be led for any distance in a boat, they should either be the special lead-cased ones made for the purpose, or they should be led through a tube of the best red rubber, so that the bilge water or any dampness can never reach them. The great difficulty of getting perfect insulation when everything is very

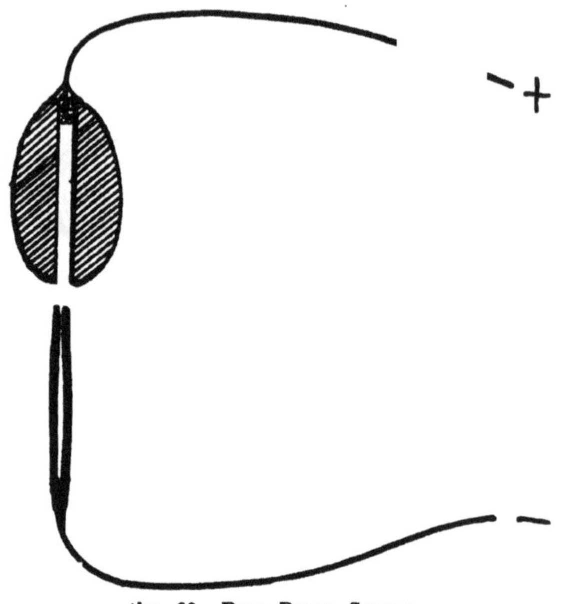

FIG. 66.—DAMP-PROOF SWITCH.

damp, is one of the difficulties that one has to contend with when afloat. The ordinary switches which answer perfectly ashore, or in motor cars, prove quite unreliable after they have been for hours in a cold, damp atmosphere, and often the battery or accumulator is found to have run down before it was expected to, owing to this cause of leakage. Any naked parts of metal in or near the terminals of the battery or accumulator should be covered with vaseline, after the wires have been connected, as this substance is a good insulator, and, moreover, prevents any corrosion from acid or sea air. Fig. 66 shows a simple, home-made substitute for the usual switch, which is practically damp-proof, and is always used by the author for sea work. One wire is soldered into a short length of brass tube, and the wire and tube well covered with a lump of gutta-percha. The other wire is soldered to a split brass pin, so that it will just enter the tube. On entering this pin into the tube and giving it a twist or two, we get a good metallic contact, and when the pin is withdrawn, and removed some distance away from the tube, there is no fear of the current creeping along the moist surface, as it does in the usual switch, where the two terminals are always so close together.

We have now to consider the sparking plug itself, which is screwed into the combustion chamber. It is made an easy fit for frequent removal, and is made gastight by means of a washer made of thin sheet copper, enclosing a thread of asbestos.

There are a great many plugs on the market, and, like carburetters, new ones are constantly coming out. They all consist of a central conductor surrounded by a good insulator, such as porcelain or mica, the outer end of the conductor being fitted with some sort of binding screw for the wire, and the inner end terminating in a short length of platinum or German silver, or other metal; a similar piece of wire is generally connected to the steel bush which screws into the metal of the engine, and the current running down the central conductor jumps across the gap, returning through the metal of the engine to the other high tension wire which is earthed to the metal of the engine.

The insulating material of the plug and the inside terminals are subjected to very great changes of temperature (as when starting an engine on a very cold day), and the pressure of the explosion is also considerable. One effect of this is that the porcelain is very liable to crack, and, however small this crack is (even if we can hardly see it) a layer of carbon will soon be deposited in it, and thus provide a short circuit for the current. Even if the porcelain be not cracked, a layer of soot or carbon will often be deposited all about the points, and a short circuit provided. This latter is especially liable to happen when we are using a mixture too rich, or when we have too much lubricating oil in the crank chamber. In the first case, as there is not enough oxygen to provide complete combustion, some of the carbon is deposited as soot, and, in the latter case, the oil works its way up, and, getting up above the piston gets burnt with the charge, and this deposits carbon.

It is always as well, before starting, just to take out the plug, and see that any soot, or carbon, or moisture, is removed. Sometimes on a cold and damp day the engine will fire once or twice and then stop, and this may be repeated several times, and is owing to the first explosion depositing moisture on the cold plug, and this moisture will short-circuit the current in the same way as the soot does. We often see the French chauffeurs on a cold day, when a start has to be made for a race, take out the plug, and heat it over a spirit lamp, so that there shall be no chance of moisture being thus deposited. It must be remembered that petrol vapour offers considerable resistance to the electric spark, and when it is compressed in

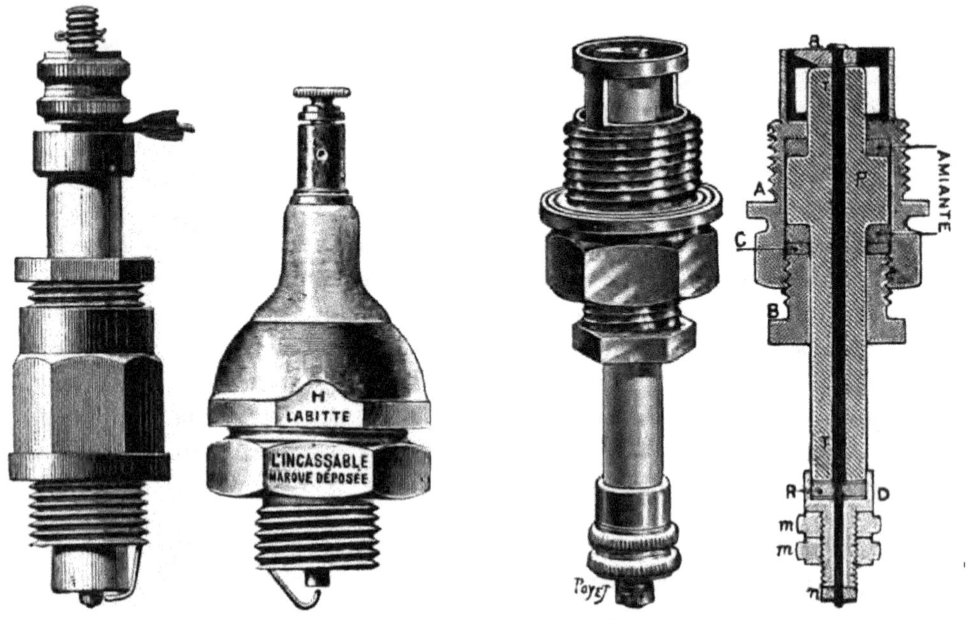

Fig. 67.—Ignition Plugs.

Fig. 68.—Ignition Plugs

the cylinder, this resistance is so high that the current will seek out any alternative route rather than jump across the sparking points through the petrol vapour.

Fig. 67 shows a De Dion plug on the left, which was one of the first to give reliable results. As the insulation is porcelain, however, there is a good deal of luck in getting a really good one, as, however carefully the porcelains are made, they are bound to differ. If we do get hold of a good porcelain these plugs will last for hundreds of hours' running; but we may have bad luck, and succeed in cracking a good many before we get a good one. The other plug in Fig. 67, and those in Fig. 68, show different forms of plugs, which may be often met with, and in Fig. 69 is shown one of the latest patterns, the E.I.C., which possesses certain distinct features of its own. The central conductor of steel is surrounded by a great many mica washers threaded on it, and tightly compressed by a nut; the outside is turned down so as to be of a conical profile, and thus makes a tight fit in the conical steel bush. When put into its place the effect of the first explosion is to drive the mica cone into the steel bush and make a gas-tight fit. After we have removed it, however, from the engine, a tap with a spanner on the outer end of the central conductor knocks it out, so that we can easily clean it.

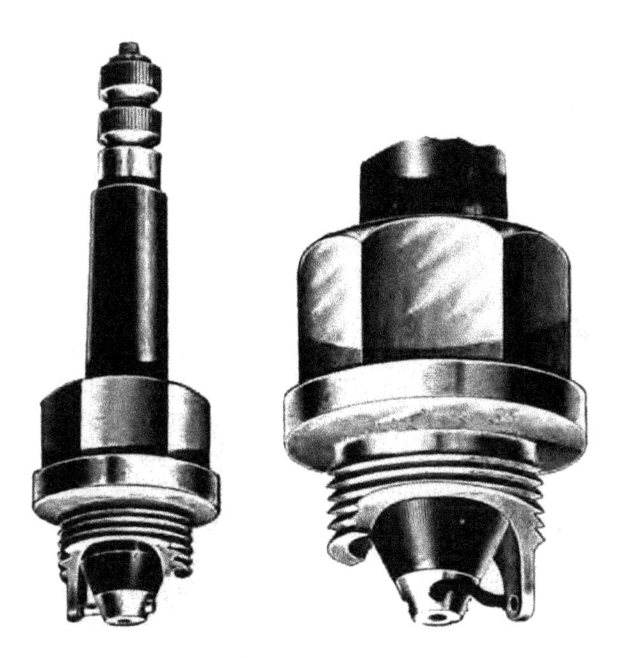

FIG. 69.—E.I.C. IGNITION PLUG.

The cross piece is made a driving fit in the central conductor; so that we can adjust it to be the proper distance from the outer bush (generally about $\frac{1}{64}$ of an inch). The author has found, in some cases, that the effect of the flame was to make a sort of a scale between this central conductor and the sliding pin (especially after it had been moved for adjustment); he found, however, by getting a piece of copper wire, of the shape shown on the large scale in Fig. 69, silver, soldered into the central conductor, so as to make a perfect electrical joint, that these plugs worked capitally, and he has them now in constant use, both afloat and ashore. The reason the wire is of the shape shown is, so that, by slightly unbending it, the proper adjustment for distance is easily obtainable.

The great advantage of these plugs is that the hotter they get the better is the insulation, and an accidental blow does not break the mica as it does the porcelain. With an engine fitted with this system of electric ignition, a set of tests should be taken, as mentioned in the case of the engine with the low tension break-spark—*i.e.*, when the engine is new and running really well, marks should be made on the flywheel to show exactly when the contact is made and broken in the primary current, when the timing gear is in a certain marked position. The voltage of the source of electricity should be taken and noted in a pocket book. In order to get the measure of the current which is being sent through the primary circuit of the coil, we must first of all stop the trembler, and to do this a small piece of wood (a piece of wooden match will do) is jammed in between the iron at the end of spring S (Fig. 64) and the ends of the soft iron rods so as to fix the trembler against the screw in the post P. We can now measure the voltage and ampèrage of the current going through the coil, seeing that the engine is in such a position as to make contact by the cam I (Fig. 65). These measurements are to be entered in our note book, and by them we can always test and see if this primary current is all right. In order to test the secondary current, we must remove the wooden wedge from the trembler, and having unscrewed the firing plug, and laid it down on some part of the engine, so that the part which screws into the combustion chamber is actually touching some clean part of the engine, we must turn the engine until the contact is made by the cam, and the trembler begins to buzz. A stream of sparks should now be visible between the sparking points of the plug, and these should be, not of a pale blue colour, but of a yellowish colour, and what is called a flaming spark.

The colour and appearance of these sparks are of the greatest importance, and, as no description will help anyone to learn to be able to judge as to this, a lesson should be taken from someone qualified to show and explain about the different forms of sparks by means of an electric installation in proper working order. If at any subsequent time we do not get the proper sort of sparks (when we know that the primary current is all right), we must look to the insulation of the secondary wire, or try a new plug; and if all this is no good, we must first of all try cleaning the platinum contacts between the trembler and the set screw, and if this does not mend matters, in all probability the coil itself is damaged, and must be repaired by a qualified electrician.

The regulation of the set screw against the trembler will alter, to some extent, the sort of sparks obtained, and when once found, the proper note given out by the trembler must be remembered, and always retained by suitable adjustment.

As we mentioned before, the compressed petrol vapour in the cylinder offers considerable resistance to the current, so much so that a poor sort of spark, which will pass freely enough when the plug is lying on the engine, will fail to jump at all when the compression is on in the cylinder.

The description of electrical ignition has involved a good deal of tedious detail about electricity and its peculiarities; but it is absolutely necessary that anyone taking charge of an oil engine thus fitted, should understand at least as much as has been described, and a careful study of this subject will more than repay anyone taking the trouble to master it.

CHAPTER IX.

SUMMARY.

EXHAUST—WATER JACKETING AND SILENCING—CONNECTIONS, JOINTS AND UNIONS—HINTS ABOUT PETROL: ITS DANGERS, STORAGE, AND DENSITY—ACETYLENE—ALCOHOL.

ONE of the commonest objections urged against the use of oil engines, both afloat and ashore, is the noise of the exhaust, and it must be confessed that, where the most modern systems of silencing the beat of the escaping gases are not used, the intermittent sharp explosions are apt to occasion considerable annoyance to those on a boat. When the steam engine first came in, it took many years to find out how to use the steam *compound*, so as to get rid of the exhaust, which, in the case of high pressure engines, is so noisy (*vide* the usual railway locomotive). No doubt, in the future, some sort of compound system will be discovered for oil engines, but at present all we can do is to let out the gases, at the end of each stroke, at a considerable pressure, and endeavour to tone down and muffle somewhat the noise thus entailed.

All oil engines should be provided with some sort of exhaust drum, or muffler, and the larger its capacity the more will it reduce the noise. It should not be less in cubic contents than about twice that of the cylinder. Better effects are produced by gradually expanding the gases in a series of exhaust drums, than by allowing them to go only into one large one before reaching the air. Fig. 70 shows a form of exhaust cooler and muffler, which can be built up of the usual pipes and fittings stocked by large pipe merchants, and as shown, it would be about right for an engine, say with two cylinders, 4 in. diameter and 6 in. stroke, running at about 800 revolutions per minute.

Such an engine would probably have an exhaust pipe of about 1½ in. diameter, and it would be better to increase this to a 2 in. pipe, so as to provide a freer exhaust for the gases. A is a vertical length of 2 in. iron water pipe, about 2 ft. long, and at B and C are fitted over it two cast iron flanges. These flanges have each a groove turned into them so as to just admit of a steel pipe 4 in. diameter being inserted into it. The ends of this steel pipe being well coated with red lead, are inserted in the grooves, and the two flanges are tightened down on the steel tube by the back nut above and the collar below. A water inlet is fitted into the steel pipe just below B, and an outlet just opposite C. The cooling water, on leaving the cylinder, is now led down into the jacket between the steel pipes and the inner 2 in. one, so as to cool down the exhaust gases which enter at A, and thus reduce their volume to a great extent.

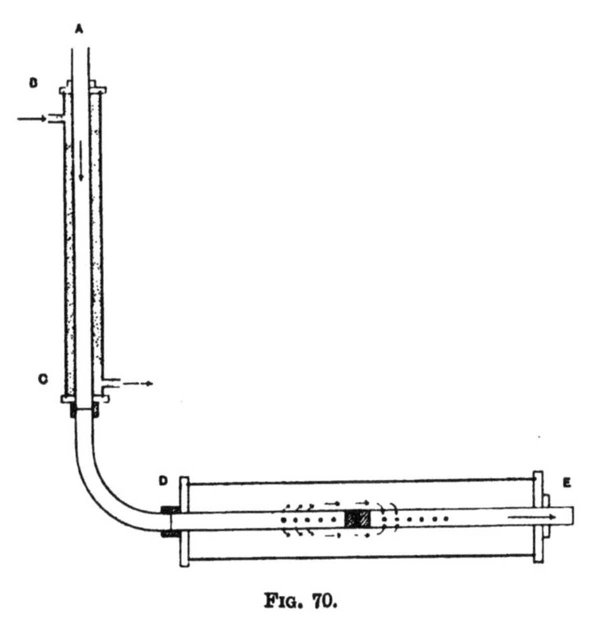

FIG. 70.

WATER JACKETED EXHAUST WITH SILENCER.

Into the lower half of the collar at C is screwed a bend, and at D is screwed another collar, and another length of 2 in. pipe. Cast iron flanges are fitted on to this lower pipe, just as in the case of the upper one, and a steel pipe of about 6 in. diameter is fixed between the two flanges, being tightened by the back nut at E. This lower 2 in. pipe is blocked up in its middle by a piece of copper hammered in and secured by a rivet, as shown, and 20 holes of ½ in. diameter are drilled in the first half of the pipe, and 24 similar holes are drilled in the second half. The incoming gases, which have been considerably cooled down by their passage through the water-cooled pipe, now enter the lower 2 in. pipe, and passing through the small holes, expand into the 6 in. steel pipe, and, having so expanded, pass down through the second series of small holes again into the 2 in. central pipe.

Another muffler exactly similar to this one, should be added on to it by suitable connections, according to the space available in the boat, but the holes in the first part of its central pipe should be 28, and there should be 32 in its second half, so as to allow the gases a more free passage after each expansion.

In some cases it is a good plan to allow the cooling water to enter the last muffler, and to pass out into the sea together with the exhaust, but in this case, care must be taken that no sea water can possibly enter by the exhaust pipe, and also that a proper fall is provided, so that the cooling water can never accumulate and block up the exhaust.

One very useful effect of proper water-jacketing is that it will prevent the very noisy explosions which are often heard in the usual exhaust drum. These latter are caused when the engine is not running properly, and misses an explosion or two; the unburnt charge then passes on to the exhaust drums, and when the engine does fire, the issuing flame will often fire the accumulated charges, and produce violent explosions in the exhaust drums. They have to be made extra thick and heavy to stand these explosions. With a proper water-jacketed pipe and drum, however, the effect of the cooling water is to kill the flame, so that any accumulated gases will not be fired. In some boats the exhaust is made to pass out under the water, but this does not do away with all noise, as a peculiar bubbling (something like that produced by boiling a bath by live steam) is produced. If the exhaust is, however, arranged this way, great care must be taken to have a small pipe open to the air, otherwise when starting the engine, or even when the engine is stopped, and begins to cool down, the sea-water will sometimes be drawn up into the exhaust drum, and may even enter the cylinder of the engine.

With water jacketed pipes and drums, a good deal of condensed water will often be found inside them, and if the way in which the drums are fitted would permit of this water accumulating anywhere, some means of being able to let it out must be provided, such as a cock or screw plug.

Except in the case of very small engines, the exhaust pipes near the engine should be of iron or steel, and not of brass or copper, as the heat is too great for the latter to last long.

One of the most important matters in connection with fitting up an oil engine in a boat, is that of the pipes, connections and joints, and a few words on them will be useful. Where petrol has to be led through a pipe, it is better to use thick lead pipe of a small bore, as it is not so likely to break where constantly bent. Copper pipe will do if the pipe has to stand any rubbing or wear, but iron pipe is not suitable for the conveyance of petrol.

As all oil engines are very liable to vibrate a good deal, when running fast, it is better to havo rubber pipes for the circulating water, than unyielding metal ones. Where metal pipes are thus used, they must be bent into a loop, and not go straight from one union to the other; the loop, of course, permits of a certain movement in the ends, without straining the metal of the pipe.

The inlet pipe for the water must always be fitted with a good strainer outside the boat, and means should always be provided for clearing out the inlet pipe from *within* the boat, should it get choked anyhow; both the inlet and outlet pipes should be so placed that they cannot be choked with mud, should the boat get ashore and lie down on her side. Strong metal cocks should always be fitted inside the boat to any pipes going through the skin; and means should always be provided for draining off all the water from the engine jackets (cylinder and exhaust) and pipes, when leaving the boat in frosty weather.

Joints and connections for petrol are troublesome to make, as this fluid will find its way out almost anywhere, and a joint that is all right for water or steam is no use for petrol. Where a joint is not often disturbed, a lead washer in the usual union will prove useful, but where a joint has often to be disturbed, it is a good plan to use a leather washer, well smeared with Sunlight Soap or fish glue in the union; a piece of this soap is very useful to carry for hurried joints or hasty stoppages of a leak, and it may not be generally known that by its use leaks and cracks in boats themselves may be stopped in a hurry, as long as the boat is in salt water; under this last condition, the soap will remain in a seam or crack for months.

To render petrol joints extra tight, it is a capital plan to smear over the outside of the joints with an excellent waterproof shellac cement made by the Waterproof Glue Co., to be obtained from Mr. R. Walmesley, Obelisk Road, Woolston, Southampton. By means of this article permanent repairs can be made of leaks and cracks in boats, in either fresh or salt water. Of course, the fewer the joints in petrol pipes the better, and where possible, pipes should be joined by soldering.

For making the joint for a union of a burner in a petrol lamp, where it gets hot, leather or lead will not do, but the author alway uses a washer of aluminium in the union, and if the two surfaces of the joint are rubbed down quite smooth on an oil stone, and the aluminium washer tightened up between them by the union, an excellent heat-proof joint is made, which can be broken as often as we wish to clean, or exchange a burner. For a heat-proof petrol joint which is a fixture, as the nipple of a burner, there is nothing better than Sellers' cement, to be obtained at 27, Canning Street, Birkenhead; and electrical ignition tubes can be mended with it. A gentle heat, after the joint has been made, sets the cement, and when set, it quite defies petrol, even under great pressure.

Petrol tanks should be filled through a funnel provided with a wire gauze strainer so as to keep out all dirt, etc. *They should never be filled quite full*, as petrol expands a good deal in warm weather, and the tank might be burst, unless provided with a small air hole, and even with this, the boat might be flooded, if the petrol expands much. Some way for the air to enter must be provided in a petrol tank, or else it will not flow down to the engine properly. This can either be by a small hole in the filling cap, or the latter must be slightly slacked before starting the engine.

Petrol vapour is heavier than air, and has therefore a tendency to accumulate in the lowest place it can settle in. Thus, if petrol is split on the floor of a room the vapour formed from it will remain for a long time spread over the floor of the room up to a certain height.

Except when the engine is running, the stop cocks in the petrol pipes should always be closed, and it is a very good thing to place these cocks where they can be seen without having to open a door or a flap. If this is done it will often be noticed that it has been forgotten to close these cocks when stopping the engine.

The stopcock fitted to the petrol tank should never be right down at the bottom of the tank, but should be about one inch above the bottom, so that any water or sediment may settle down in the tank, and not run down to the carburetter. Iron pipes are not suitable for the conveyance of petrol as they generally contain scale, which may get loose and cause trouble. Great care should always be taken to see that the pipe does not form an *upward* arch anywhere between the tank and the carburetter; if it does, air will accumulate there, and the pipe will be *air-locked*, and the petrol will not flow freely.

It can hardly be too frequently urged on oil engineers never to use any oil for lubricating the cylinder except special cylinder oil for oil engines. Where two sorts of oil are used, that for the cylinder should be kept in a can of *different shape* to the others. A different colour will not show on a

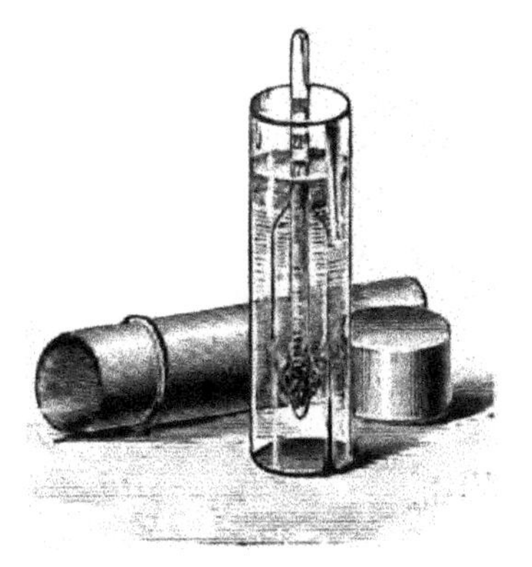

FIG. 71.
DENSIMETER.

dark night. The author has only one sort of oil on board so as to avoid all mistakes, and this cylinder oil and some lubricating grease will provide for all requirements. Engines with air-cooled cylinders require a different oil to those with water-cooled cylinders.

Whenever any petrol is bought in any out-of-the-way place, it is just as well to test its density or specific gravity, and this is done with a densimeter, shown in Fig. 71. It consists of a small glass bottle-shaped float, ballasted with small shot, so that when immersed in petrol of a correct density, at a temperature of 60° F., it will just sink down so that the mark ·680 is just awash. If the petrol is of a heavier nature the float will rise so that a higher number than ·680 is now awash, and if the instrument were immersed in a lighter fluid, say what we call here gasoline, it would sink down below the former marks. In graduating these instruments the density of water is taken as 1, so that as one gallon of water weighs 10 lbs., one gallon of petrol weighs ·68 lbs.

The petrol as issued by the leading firms in England has a density of ·680 at 60° F., and this density should be insisted on; always remembering that if a sample of proper petrol were taken at a temperature 50° F., it would show a density of ·685, and if its temperature were 70° F., its density would be diminished to ·675. The allowance for temperature, therefore, is ·005 for each 10 degrees F.

No greater quantity that 60 gallons of petrol may be kept by anyone, except by special permission, and the fluid must be in separate vessels of an approved pattern, each vessel to hold only two gallons.

In case a mass of petrol catches fire, it is of no use to pour water on it, as the spirit floats on the water and goes on burning. Sand or earth will smother the flame, but as we cannot well carry about either of these in a boat it is advisable to have some strong *liquor ammoniæ*, kept in a proper bottle or jar. Some of this poured on to a burning mass of petrol in a more or less confined space will put out the flame. The fumes of the ammonia should not be inhaled as they are deleterious.

In starting an oil engine the handle of the flywheel should never be held with a rigid arm. Always keep the arm bent, and use the muscles of the elbow and shoulder to turn the wheel. The thumb also should be never under the handle, but on the top—*i.e.*, the same side as the fingers are held. If this is done, we shall never feel the back kick of the flywheel, which sometimes happens and rather alarms beginners.

Before finally dismissing the subject of internal combustion motors, for those which burn the fuel in other ways than in the cylinder, we must say a few words on two modern rivals to petrol—*i.e.*, acetylene gas and alcohol.

Acetylene gas as is now generally known, is formed by simply adding carbide of calcium to water, and when it was first brought out in a mercantile form a few years back great expectations were raised as to its possible use in internal combustion motors, and many experiments have been made in this direction.

Compared with coal gas, acetylene produces a far greater heat and power, bulk for bulk, but when the cost and the bulk of the necessary apparatus and the carbide is compared with petrol it will be seen that the latter still holds the field for marine motors. The wholesale price for carbide is now about £20 per ton, and each pound weight of it should produce about 5 cubic feet of acetylene gas, and we should not get much more than a half horse power per hour out of these 5 cubic feet, whereas half a pint of petrol will do this.

The French Government have now, for some time, been trying to see if alcohol could not be used for motors in place of petrol, because, if successful, it would create a new future for the beet-root industry, which has for so long been artificially maintained by the bounty system. So far, however, these experiments have not proved that alcohol can be considered a serious rival to petrol. There is a great difficulty in getting the exact amount of air in the carburetter with alcohol, and, if this be not done, the combustion leaves certain corrosive products in the cylinder, which would soon ruin the motor. A certain amount of success has been obtained by suitable mixtures of petrol and alcohol, when used in a special type of motor with high initial compression, and a long stroke.

Lately some of the most daring racing chauffeurs have obtained startling results (in two senses) by mixing the explosive picric acid with petrol, but these experiments have as yet not reached the practical stage. The author has no doubt that some safe substance will soon be found to mix with petrol, and thus enable more power to be obtained out of it than can at present be got; and this would be a great boon to users afloat; especially if, at the same time, the new combination fuel were rendered still safer in use.

If at any time an owner of a motor boat should find himself stranded in some out-of-the-way place with his petrol tank empty, and none procurable, it is worth while to know that most petrol motors will work fairly well with benzoline, and there are not many places now where one cannot obtain the latter spirit, because it is used by plumbers and painters so much in the spirit lamps for soldering and burning off paint. Chemists often also stock benzoline, as it is used for cleaning purposes. All jet carburetters will not work with benzoline, though all surface ones will do so, as considerably more of this fluid has to be used than in the case of petrol, but an ingenious man will often find out how to give the extra quantity of fluid by some means (such as increased pressure, for instance, in the tank).

Methylated spirits of wine will also sometimes work a petrol motor quite well enough to get home should the supply of petrol run out.

CHAPTER X.

SUMMARY.

FIVE TON AUXILIARY SAILING BOAT ON IMPROVED SAFETY SYSTEM—ADVANTAGES OF AUXILIARY POWER—PROPOSED POWER—WINDLASS FOR OIL ENGINES.

THE drawings shewn on this and following pages herewith show a system which has been designed by the author for providing any small boat, from a 10 ft. yacht's punt to an open or half-decked fishing boat of about 17 feet long, with a compact and portable oil engine, and the following points have been kept in view, and may, it is thought, be fairly claimed for the system.

(1) SAFETY.—Owing to the fact that there is never at any time any petrol in the actual boat itself.

(2) RELIABILITY.—Owing to the way in which the tank and carburetter are fitted, so as not to be affected by the motion of the waves or the pressure of a sail.

(3) BETTER POSITION OF THE ENGINE.—Owing to the position, it is above any water in the boat, even if half full; and, owing to the way in which it is connected to the screw shaft, the engine can be run at the proper speed to get power, the screw only running at a slower speed so as to get efficiency.

(4) FACILITY OF GETTING AT AND REMOVING ENGINE.—Owing to the fewness and simplicity of the connections of engine it can be removed very quickly, either for storing it ashore, or for rendering it easier to hoist up the boat in davits or elswhere.

The full-page drawing Fig. 75 shows a 16 ft. fishing boat, which may be either open or half-decked, and the figures 72, 73 and 74 show the various details of fitting the engine and its accessories, and it will be seen that (taking the boat as a 15 ft. 6 in. one, which it really is) the whole of this boat, except 18 in. of its after end is available for the owner.

Taking the engine first, it is a very difficult matter to choose a motor from among the very many which are now on the market, but a bicycle motor is suggested; because that if well and durably made, the greatest

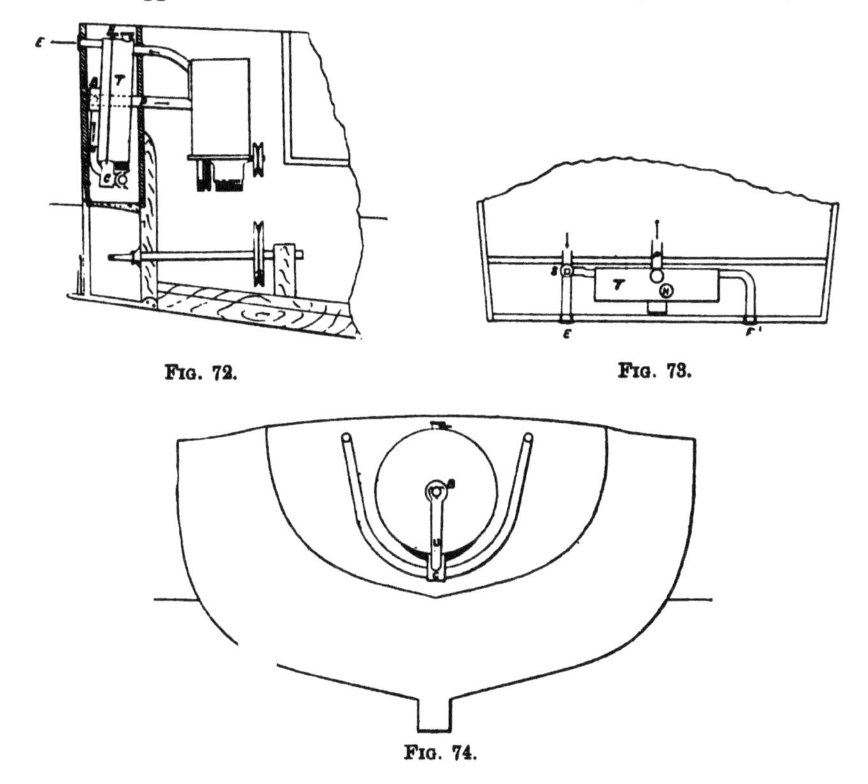

FIG. 72. FIG. 73.

FIG. 74.

power for a given weight is obtainable with this form, and that the best made ones are really durable is becoming more evident every month. The author is in possession of a French one which has now done a good two years' running on a bicycle without any serious sign of wear, and he knows of a similar one which has run 7,000 miles on the road, and the only main portion showing wear was bearing at one end of the engine shaft, where the pull of the belt came on it, and which had to be replaced. So much for a well made, air cooled, small motor!

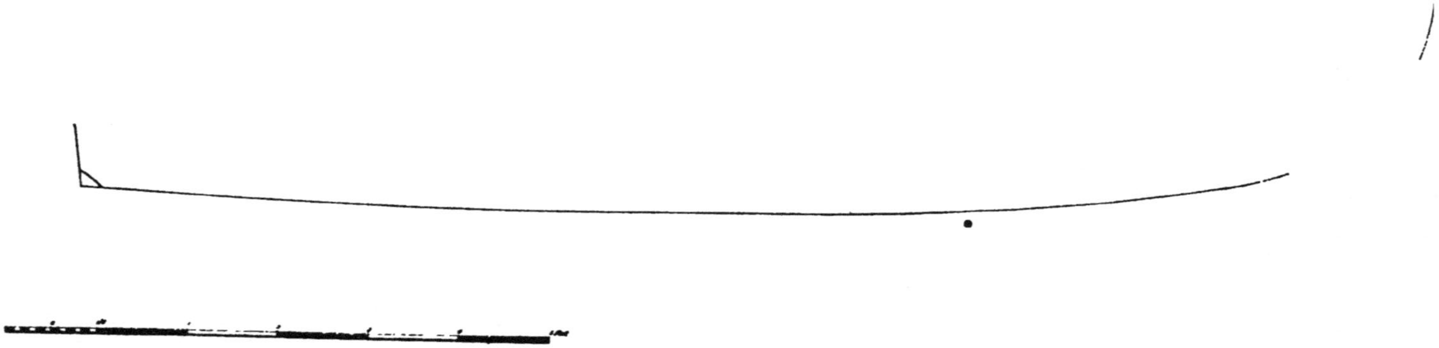

Fig. 75.

16-ft. Auxiliary Fishing Boat.

Of the various bicycle motors, the Simms has been selected, not only because it is fitted with an admirable system of air-cooling by a small fan, and, being made in England, spare parts can easily be replaced, also because it is entirely self-contained, with its magneto system of ignition, so that there are no wires or batteries to connect up when shipping or unshipping it.

This engine has been already illustrated in former chapters, and it only remains to be said that by means of a groove in the flywheel, and a belt, a small fan is driven at a high speed by the engine; and this fan is used to *exhaust* the air from a brass case which fits all round that part of the engine above the bed plate. The case is made an airtight fit on to the bedplate, so that the only way for the outside air to run in and replace the air withdrawn by the action of the fan is through certain holes. By placing these holes suitably all round the engine, we can thus get effective air-cooling just where we want it, and the consequence is that the engine is quite independent of the wind, and can be run for hours, either with or against it. Fig. 72 shows the engine in its case.

The engine can be run up to over 2,000 revolutions per minute, but as it is advisable not to continually run it as hard as it can be made to go, the speed that will be suggested is 1,400 per minute; which, allowing for the power absorbed by the fan, should enable it to give out about 1½ h.p. By means of a rope or belt drive, the speed of the screw can be kept at about 700, which is quite fast enough for efficiency. The rope gives quiet running, and by keeping the rope slack, and tightening it by a jockey pulley, a very simple mean of stopping or starting is provided. No stern gear is suggested, as in a small boat such is realy not wanted, although it could be added by having double pulleys, and reeving a simple rope on one set, and a crossed rope on the other set, with jockey pulleys to tighten either at will.

It is a great gain thus to get rid of all water pipes, pumps and connections, especially when it is required to unship the engine at times.

So much for the engine, and we will now consider seriatim the advantages claimed for the system generally.

(1) SAFETY.—The boat itself is really a 15 ft. 6 in. one, but a false transom is added, 6 in. abaft the ordinary one, and the chamber thus formed between them is lined with thin sheet copper, so as to be oil tight,

and it is used for the tank and carburetter, *so that at no time whetever is there a single drop of petrol inside the boat proper.* The engine draws in, at each suction stroke, its mixture from the carburetter through the pipe P. This pipe is made an air-tight fit through the transom, and rests against the false transom aft. On this pipe the cylindrical tank T is fitted so as to revolve round it freely in any direction, as the tank is constructed with a piece of tube across its centre, which tube just fits over the pipe P, At the bottom of the tank is screwed the carburetter C, which is of the spraying valve type (*vide* Fig 44), and the engine draws in its charge through the carburetter C (where it becomes carburetted), up through the pipe U to the box B, which surrounds the pipe P, and revolves round it together with the tank. Holes are pierced in the pipe P just inside the box B, so that the mixture enters the pipe P thereby and goes on to the engine.

By thus keeping all petrol out of the boat, not only do we get practically safety, but we avoid all the trouble which often occurs when the spraying valve of the carburetter leaks, and thus floods the engine. With the ordinary fitting, when this occurs, we have to go through a spell of working the engine by hand in order to get rid of the extra petrol, and with the usual two-cycle engine fitting, the petrol often floods the crank chamber, in which case, if there is any considerable amount of petrol there, we may have to let it run out, and thus waste all our lubricating oil.

Petrol is very liable to get through the spraying valve, but if this happens in the boat we are discussing, it only runs into the chamber aft, where it can do no harm, and moreover is *visible*, so that we can easily let it out, by opening a small cock, which lets it run into the sea.

The tank shown would run the engins for about ten hours, and spare petrol could be carried in wedge-shape tanks on either side of the circular tank.

(2) Reliability.—Absolute reliability is probably unobtainable at sea with an oil engine, but it can be more or less approached. One great source of trouble is often that in a sea way, or especially when the boat is listed over a good deal by sail pressure, the valve of the carburetter, or its float feed, does not work regulary. The valves and feed are generally made and designed to work in a vertical position, and when they are at an angle they refuse to work evenly. Another point is that the valve having been set to work properly for the pressure of the petrol coming from the tank, will not

work properly when the boat is listed over, because the surface of the petrol in the tank is not so high above the valve as it was when the boat was on an even keel, and thus the pressure is reduced.

The patent tank and carburetter, shown in the detail drawings 72, 73 and 74 were designed by the writer to overcome these defects, and it will be seen that the weight of the carburetter, assisted by some added lead, is sufficient to keep it not only in a vertical position, but always the same distance below the surface of the petrol in the tank, independently of the position of the boat.

The feed of petrol from the tank is regulated by the needle valve worked by the milled head H, which should be graduated, so as to be able always to set it again at what has been found to give good results. Should the carburetter need seeing to, the tank is turned upside down, and the carburetter can be easily unscrewed from the tank and replaced again in a minute.

One of the most essential points for regular working is to get the right amount of heated air for the carburetter, and this will vary according to the weather. This is provided for in the design by causing the exhaust from the engine to pass out through the pipe E. A stop-cock is fitted at S, so that a certain portion of the hot exhaust can be passed through the alternative exhaust E_1, through the bent pipe below the tank. This pipe is bent into a circle, so that whatever the position of the tank and carburetter is, the latter will always take its heated air from the surface of the hot pipe, and as the temperature of this pipe can be regulated by the stop cock, we can always thus ensure the best results in its working.

(3) Better Position of Engine.—There are several great objections to fitting the engine in a small boat in the usual position, more or less down in the bottom of the boat. First of all, when lying at moorings the risk of its getting drowned, either by salt or fresh water, is a continual source of anxiety to the owner and necessitates someone being always handy to pump out the boat whenever water is accumulating either from a leak, spray or rain.

Another point is that we want to get the screw shaft down as low as possible, and more or less horizontal. Now, when the engine is connected direct to the tail shaft, we often have to tilt it backwards at a considerable angle, in order that the flywheel shall clear the bottom of the boat. This practice not only leads to considerable loss of power, but it also leads to difficulties in lubricating the engine, which (as has been found by experience in motor bicycles) ought to work in a vertical position. When the boat begins to move, and squats down aft, these two evils are exaggerated.

A third point is that, in order to get the proper power out of a small motor, as a bicycle one, it must run at a considerable speed, and this speed is too great to get the greatest efficiency out of the propeller.

In the position shown, however, all these difficulties are got over, and a simple, quiet and cheap means of gearing down the revolutions of the engine is provided by the rope drive, which also replaces the usual clutch for starting the engine free from the propeller.

(4) Facility of Getting at and Removing Engine.—Messing about down in the bottom of the boat is not a pleasant job when anything has to be inspected or adjusted, and the position chosen in the design not only permits of being able to see much easier what is wrong, but also to make any adjustment with greater expedition and comfort. As regards removing the engine altogether, there are probably many men who would much like to have a motor in their small fishing boat, or tender for a small yacht; and the former perhaps often like to leave their boat for many days together unattended at moorings, and perhaps only use her at week-ends. Now the proper place for the engine is ashore in a boathouse, and not in such a boat; and, as regards the small yacht's tender, the being able to easily remove the engine renders the hoisting up of the boat a much easier job. It makes a great difference in the running and life of an engine if we periodically give it a thorough cleaning out with paraffin and petrol so as to remove soot and carbon, &c. Now, this is a dirty job in a boat, and, consequently is often shirked, but if we can take the engine ashore anywhere, it becomes a comparatively simple matter to do this and assist our engine in its future work.

In order to remove our engine, the only connections to touch are: first, to pull off the rubber suction pipe from the inlet. Secondly, to unscrew the union of the flexible exhaust pipe. Thirdly, to loosen four nuts on the holding down bolts, which latter being hinged, open outwards, and the loose rope having been unhooked off the driving pulley, the engine, with its case, is lifted up, and carried ashore by a handle at the top, and the total weight to be thus carried, is just over 56 lbs. There are, of course, no electric wires, or pump connection to bother us.

As regards the question of putting weight in the end of a small boat, it must be remembered that, in such a craft, the weight of a man is generally on the stern seat, and the average weight of a person is more than the whole of the engine and gear described; and when the latter are fitted the person will sit more forward, so that the boat's liveliness in a seaway would not be interfered with.

It only remains to be added that the false transom could be added to existing boats, as well as built into new ones.

The drawings (76 and 77) and description on this and following pages show an oil engine as applied to a larger type of boat than the small one previously described—*i.e.*, a 5 ton sailing boat belonging to the author—and, as the system has now been tried in all sorts of weather, and at all times of the year during the past two years, and the boat has proved an unqualified success, he has no hesitation in recommending anyone desirous of fitting an oil engine into a sailing boat as an auxiliary to study the system by which practically the risk of using petrol afloat in a decked boat has beed avoided.

As in the case of the smaller type of auxiliary and yachts tender, ***not a drop of petrol is ever allowed inside the boat***, but it is kept in suitable receptacles on the deck, and the engine draws in its charge from the deck straight into the cylinder.

To begin with the boat itself, illustrated in Figs. 76 and 77, is 26 ft. 5 in. length over all, 21 ft. on w.l., and 7 ft. 2 in. beam, and draws 3 ft. 6 in. She is half decked, with a cockpit and a small cabin, and carries 27 cwt. of lead. Her lines were a modification of some published in *The Yachtsman* some years ago, and, as a sailing boat, and especially as a sea boat, she proved so useful that no less than six boats have been built on her original moulds. Fig. 78 shows her midship section.

The author was very desirous of putting an oil engine into her, and much prefered petrol, but did not care to have any in the boat; and having bought a French motor car using petrol, fitted it in the position shown in the diagram. The engine is a 4 h.p. single cylinder one, of 90 millimetres diameter, and 110 stroke, and runs at about 700 per minute. Its consumption is about three pints of petrol per hour.

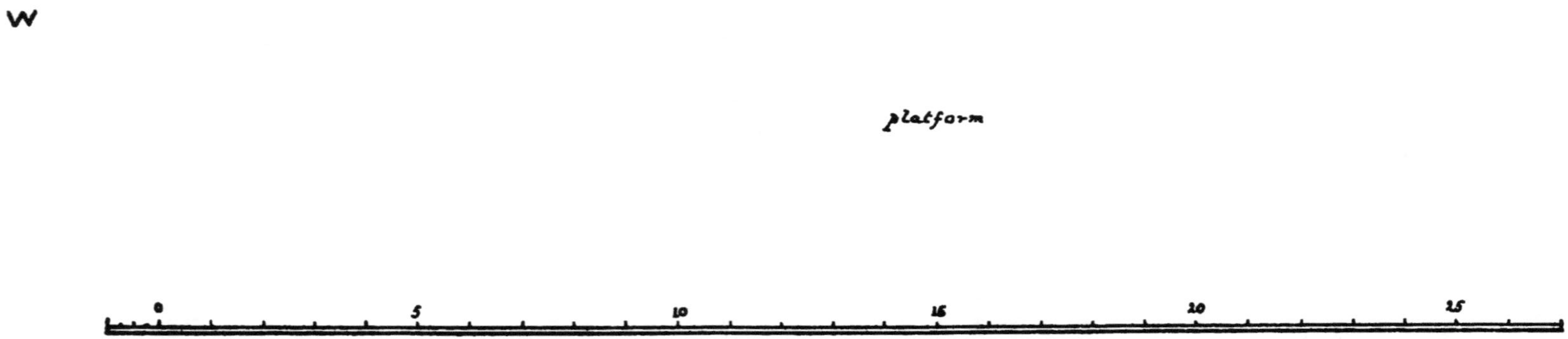

W

platform

0 5 10 15 20 25

Fig. 76.

Five-Ton Sailing Boat Fitted with Motor.

Section Shewing Machinery and Pipe Connectoins.

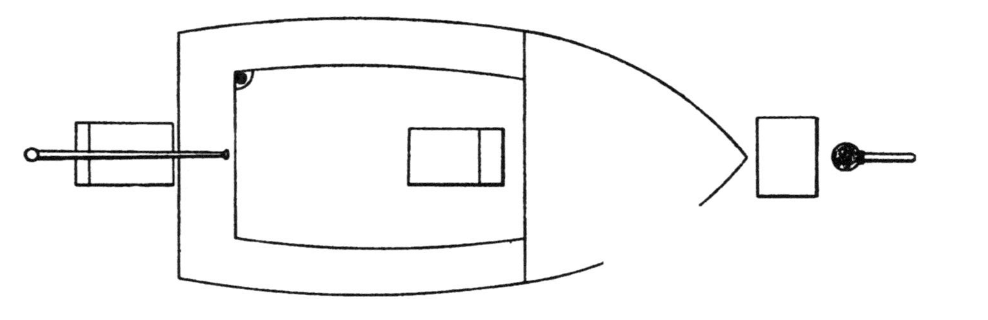

Fig. 77.—Plans of Five-ton Sailing Boat Fitted with Motor.

It is started by a handle working a bicycle chain sprocket wheel near the top of the engine, the chain going down round a free-wheel clutch keyed on the engine shaft. As soon as the engine starts it runs *free-wheel*, and the chain and upper wheel are stationary.

Just under the tiller on the deck is a teak case B, which contains the carburetter, a supply of petrol for 1¼ hours' run, and a small heater for the air entering the carburetter, which is warmed by passing some of the exhaust through it. The engine draws each charge of mixture from the carburetter down through a 1¼ in. compo. pipe V, and the exhaust passes through a water-jacketed bent pipe to the drum D, whence the main portion of it passes up through the pipe E on the fore side of the mast, and the remaining portion of it passes up into the case B, and having warmed the air for the carburetter, passes out through the pipe E, just under the counter. Just abaft the mast on the deck is a teak case A, which contains a copper tank holding three gallons of petrol, and a small copper pipe runs

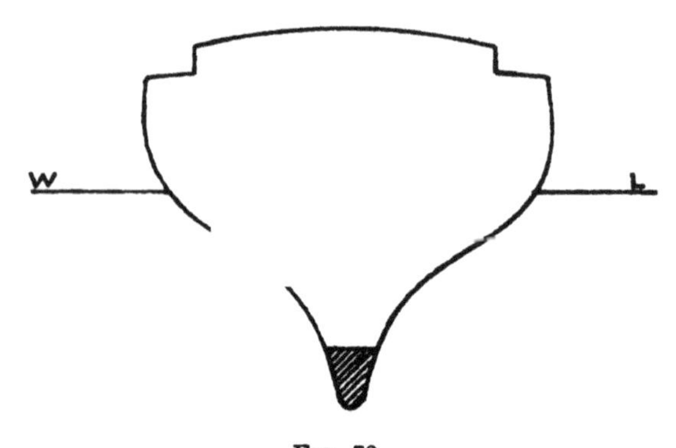

Fig. 78.
Midship Section.

along the deck close to and secured to the coamings of the well, which protect it from injury, and it feeds the tank in the case B whenever a stop valve is opened.

Now it will be seen that not a single drop of petrol can ever enter the boat, as the engine only draws in the vapour from B, and the pipe V is bent up so as to be above the level of the petrol tank in it, so that, even if the carburetter were to be flooded with petrol, it could not be drawn into the engine.

Independently of the safety thus gained, the working, and particularly the starting of the engine is much simplified, because whenever an engine gets flooded with petrol, it is a very tedious job to get it clear again, and until this is done, the engine will not start. When the spraying valve affixed to the crank chamber of a two-cycle engine leaks (as they often will), the petrol either runs down into the boat, where it may constitute a serious danger, or into the crank chamber; and, in the latter case, the crank chamber has sometimes to be drained out, and all the lubricating oil therein wasted. By having the spraying valve in a separate case on deck, any leakage being visible is at once detected, and the waste petrol easily run into the sea. The amount of exhaust passing through B to warm the air is regulated by a cock, so that, in winter, more can be thus used than in summer. In fitting an exhaust pipe to discharge under the counter, great care has to be taken that no water can possibly enter and run down to the exhaust drum or engine, when running before a following sea; this is avoided by the pipe E coming up above the deck into the case B. Owing to the fact that the exhaust being first water-jacketted, and then having two escapes open to it, there is hardly any noise, and the little there is from the forward pipe is further muffled by the mast being between it and anyone in the well.

As regards the room taken up in the boat by the machinery and its accessories, the part of the deck just abaft the mast, and under the tiller, was chosen as being of no other use, and the only part of the engine above the platform is in a teak case 12 in. wide and 10 in. high, so that it forms a useful footrest or seat.

The propeller is a plain one, and a small friction clutch just abaft the engine permits of its being disconnected from the engine. The ignition is electric (jump spark). The two tanks will run the engine full speed for about 9 hours, and a speed of just over four knots is obtained.

The owner has now fully tested the boat in all sorts of weather, during the summer and winter, and (bar the usual troubles attendant on electrical ignition), never has any trouble with the engine; and he has no hesitation in going out trawling at night with a stove going in the cabin for cooking and warmth, which makes all the difference in fishing during the winter.

No trouble or damage has ever happened with or to the deck fittings and pipe, the tanks being contained in strong teak cases, and the pipe being quite protected from injury by running along the corner between the coaming and the deck.

It is difficult to exaggerate the tremendous advantage gained by having an engine in a boat always ready, to be started in a few seconds. If there is a long thrash to windward, the engine is kept going all the time. If there is a doubt about being able to weather a point or pier, it is easier to start the engine and do so, than to make a tack out into a foul tide. In trawling, an extra heave or two can be gained in an afternoon's fishing by having the engine to help us back over the foul tide; and, even if towing the trawl by the engine be forbidden by any local regulations, in light weather, when the trawl hangs up, a turn or two of the screw will nearly always start it off again. It is often very desirable to make a jibe or two when trawling, so as to clear foul ground, and this is often a very tedious process with a sailing boat; but again, in this case, a few turns of the screw will turn the boat round enough for the mainsail to jibe over. These are some of the advantages gained, beyond the very obvious one of being able to get home in a calm, and of entering a harbour channel when the wind will not let us lay the course.

The boat, as shown, is fitted with a large cockpit, and only a small cabin, as she was meant for day work, but with only a small steering cockpit a capital cabin could be obtained, and, with all the advantages described above, it would not be easy to find a more suitable type for a small single-handed cruiser, as she possesses excellent sea-going qualities, combined with only 3 ft. 6 in. draught.

When we have gone to the trouble and expense of fitting up an engine in our boat, it is just as well to be able to call on it not only for propulsion, but for other hard work, and in Fig. 3, *et seq.*, a proposed plan is shown for using the engine for a windlass which would be extremely useful for weighing anchor, hauling off a kedge, or getting a trawl on board, or hoisting the sails.

There are probably many men who use the sea, who have all the tastes and qualifications in them for making first-rate single-handed sailors, and who often long to do so, but are afraid of the hard, physical work entailed at times; but, to such men, the oil engine opens up a new field, because it will exactly supply the power when and where wanted. The skilful use of machinery is in no way antagonistic to a proper love of the sea, as witness the wonderful mechanical contrivances utilised by that prince of single-handed sailors, the late Lord Dufferin, in his boat *Lady Hermione.* But, unlike Captain Slocum, or that inimitable yachtsman,

"Diogenes, jun." we do not all possess the valuable gifts of physical strength and endurance, and where these gifts are lacking, then intelligently used mechanism comes in.

Referring again to our windlass, Fig. 79 shows it as fitted on the deck of our boat, and driven from the flywheel of the engine by means of a

W

FIG. 79.
OIL ENGINE DRIVING WINDLASS.

flexible shaft, which can be taken off when not wanted. Fig. 80 shows the windlass, and it consists of two oak cheeks, going down through the deck to the keel. Between these cheeks is a cast iron wheel fitted on to a horizontal shaft, which works in metal bearings affixed to the after edges of

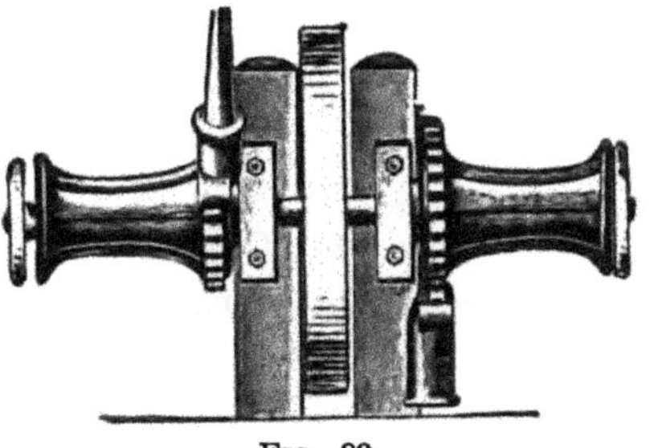

FIG. 80.
WINDLASS.

the oak cheeks, as show. The cast iron wheel is made with teeth which engage with a worm, coming up through the deck at an angle as shown, and this worm and the wheel are covered in with a case to keep the lubricating grease on them.

It will be seen now, that if we cause the worm to revolve, the worm wheel will also revolve, and will carry round the horizontal iron shaft. On each end of this shaft is a barrel which is free to revolve round the shaft, except when the hand wheel (shown at either end) is tightened so as to jam a cone into the end of the barrel. When this is done, the revolving iron shaft will carry round with it either one or both barrels, according as one or both hand-wheels are tightened. Now, supposing both the hand-wheels to be loose, we have an ordinary windlass, and either barrel can be worked by hand by means of levers inserted into the usual sockets shown in the drawing. Pawls fitted on the cheeks prevent the barrels from turning backwards. If, however, we cause the worm to revolve and tighten up one or both hand-wheels, we have at once a powerful mechanical windlass ready for any purposes required.

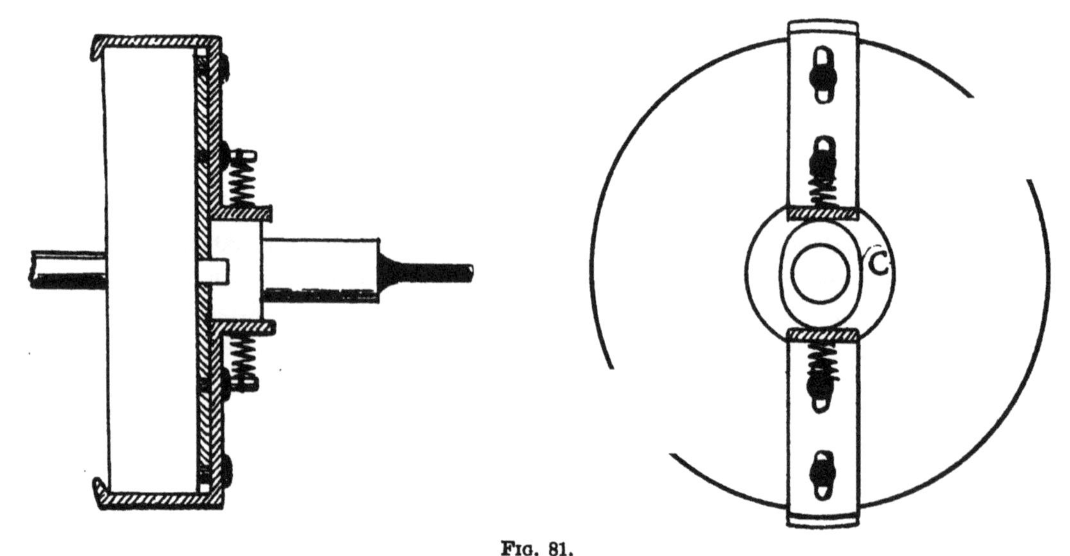

Fig. 81.

Friction Grip for Flywheel.

The chief difficulty in fitting an oil engine to drive anything like a windlass is that, if the work is to heavy for the engine, it will slow down and stop. A steam engine, on the contrary, will hold on and wait till it can get another pull when it is overloaded, as it is an elastic force, but an oil engine will not work thus, so we must provide some means of easing it when overloaded.

Fig. 81 shows how this is obtained. The flexible shaft terminates in a cross piece of sheet steel, the length of which is a little less than the diameter of the flywheel. On this cross piece are mounted two gripping pieces of sheet steel, free to slide a little in and out, so that they can be hooked over the edge of the flywheel, and when their two spiral springs tighten them, so as to form a sort of brake on the circumference of the flywheel. The gripping pieces are shod with leather where they grip. In order to open and close these gripping pieces, an oval cam, shown as C, Fig. 81, is fitted on to a collar, which can be turned round on the end of the flexible shaft by hand. When this cam is in the position shown, the two gripping pieces will just slip over the flywheel, and they are kept central by means of a hole in the centre, into which the end of the engine shaft projecting from the flywheel just enters. When thus shipped, if we turn the cam round a half circle the spiral springs will pull the two gripping pieces together and thus put the brake on the flywheel. If the tension of these springs is so adjusted as to allow the flywheel to slip when the engine is overloaded, we shall be able to keep the engine running as long as we like if the windlass is revolving or not.

It will be only a matter of a trial or two to adjust the springs so as to overcome the centrifugal force, and put just the proper strain on the gripping pieces.

When we wish to use the windlass by power, therefore, we first of all ship the forward end of the flexible shaft on to the end of the worm which comes down through the deck, and then start our engine; as soon as it runs well, we ship the after end of the flexible shaft on to the flywheel, and as soon as the oval cam is turned, the flywheel is gripped, and the windlass revolved as long as we want it. Suppose we are getting the anchor, as soon as it is up to the hawse pipe, we stop the barrel from revolving by slackening the hand wheel, and the barrel is held by the pawl, so that the anchor cannot lower itself away, we then go to the engine, and by slightly gripping by hand the collar on the shaft which works the oval cam, this latter opens the gripping pieces, so that they no longer grip the flywheel, and the shaft stops revolving. It can then be unshipped from the flywheel, and the engine can be at once used for the propeller without having stopped it at all.

The forward end of the flexible shaft is fitted with a sort of clock key, which ships over the square end of the worm, which projects downwards through the deck.

The flexible shaft will be either Stow's wire rope shaft, or the jointed telescopic one shown in Fig. 82.

In designing this windlass with its gear, the author has endeavoured to keep all mechanical details as simple as possible, so that any intelligent ship engineers could make and fix it into a boat. Some sort of windlass will be required in every boat, and it should not be a matter of great expense to add the mechanism shown.

The flexible shaft will probably be expensive, and it may be worth while just mentioning an experiment which the author made some years ago, in order to convey the power of a small oil engine on to a shaft.

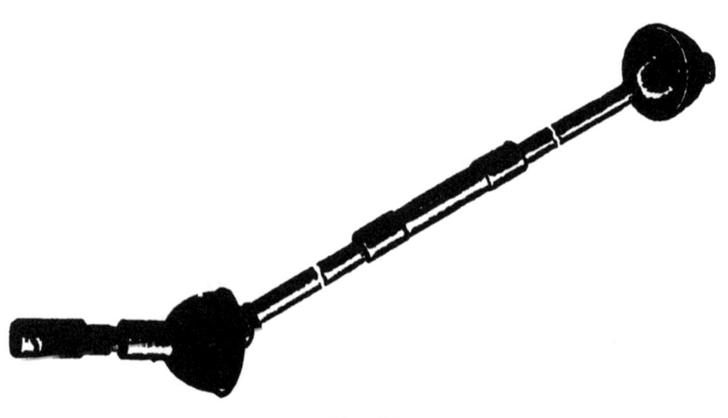

Fig. 82.
Flexible Telescopic Shaft.

If a canvas and rubber pipe of a fair diameter, and of the best quality, be taken, and it be rammed tight with bottle corks which just fit it, with plenty of blacklead all about the corks inside the tube, and if the proper metal connections be tightly fixed into the ends of the tube with wire seizings, as long as the tube is kept round, and prevented from crippling itself inwards, it is surprising what power it will convey; and as the corks will prevent this crippling, and both they and the inside of the tube soon get burnished with the blacklead, the whole apparatus should last for a good time. It will be quite flexible enough for such a purpose as we are describing.

CHAPTER XI.

SUMMARY.

HINTS FOR FITTING OUT AND LAYING UP MOTOR BOATS—ENGINE REPAIRS AND ADJUSTMENTS.

ANYTHING of a mechanical nature, be it a motor, or a gun, should always be put away for the slack season in a proper state of cleanliness, and it is far better to do any necessary adjustments before laying up than just before fitting out, not only because it is easier then to remember any small things which may require adjustment, but it is far better for an owner of a motor boat to know that the motor, at all events, is all ready for another season's run, and he can then devote all his attention and time to the fitting out of the boat itself, knowing that when this is done, and the motor properly provided with fuel and electricity it will go right away, and require no more thought.

In the case of steam launch or yacht, there is sure to be an engineer, and he will take care that the engine is put away in a proper state, not only for the trouble that it will save him next season, but also, perhaps, because he will not be inclined to hurry over the laying up of the engine, knowing that, as soon as this is done, he will probably come on to a lower rate of wages.

Of course, the great advantage of an oil engine is that it does not necessitate the use of an engineer, so the owner often takes charge of it himself, and, unfortunately for the engine, this often means that it is left just as it last stopped, and put away with the boat for the slack season.

The following remarks as to fitting out, therefore, are supposed to apply to an engine which has been so left, but it is hoped that anyone reading them will carry out the hints here given when next laying up his boat for the winter.

The first thing to do will be to remember if any adjustments were considered necessary before laying up. Was there any particular knock in the cylinder, or any play observable in the shaft bearings? Did any of the bearings ever show signs of heating after a lengthy run? Did the friction clutch ever show signs of slipping, or did the reversible blades of the screw ever jam, or refuse to work, or are the latter now too loose in their bearings?

If any of the foregoing were ever noted in the last season's running, now is the time to rectify matters; it is a fatal mistake to say: "Oh, she will go along for another season!" Probably just in the middle of it, when we want to use our boat every day, we shall have to waste perhaps a week or more in getting matters put right.

The first thing to do will be to take off the cylinder cover and remove the cylinder from the crank chamber by pulling it up over the piston, which will remain sticking up out of the crank chamber. We can then feel if the lower end of the connecting rod has worked loose on the crank-pin, and if this is the case, the brasses must be tightened, but not too much so; and care taken that the oil holes and channels are free for the passage of the lubricating oil. If the engine shaft shows any play in its bearings, then it will have to be unshipped, and a new collar turned to make a better fit on the shaft. This latter, of course, will be an enginneer's job. If, however, there is not much play in the engine shaft, and we have properly tightened up the crank brasses, there will be no need to disturb the lower part of the engine at all, and the crank chamber can be well cleaned out by means of plenty of waste and paraffin oil, the final cleaning to be done with petrol. There will be more or less hard soot on the top of the piston, which must be scraped off with an old knife, and the piston rings must be worked quite loose and free in their places, by means of paraffin oil and petrol, and the rings must be turned round so that the joints in them are not over one another, but at equal distances apart. We must now feel if the gudgeon pin has worked loose at all in the piston, and, if this has happened, the set screws must be tightened up so as to stop this. The upper bearing of the connecting rod on the gudgeon pin is very likely show signs of wear, because this is the most difficult part of an oil engine to lubricate properly; any slack here should be taken up, and if the gudgeon pin is badly cut or worn a new one must be fitted.

The cylinder having been well cleaned inside, can now be carefully slipped down over the piston, and on to its seat on the crank case. The cylinder head and valve passages will be found to have more or less hard soot on them, and this must be carefully scraped off and removed. The valves must be removed and carefully examined. If the stem of the exhaust valve shows much signs of being burnt, never trust it, but put in a new one. The seatings of both inlet and exhaust valves must be examined, and the valves ground in with oilstone dust and water until they show a good even bearing all round. As soon as this is obtained, carefully remove all the oilstone dust and burnish the seatings by turning round and round with paraffin oil as a lubricant. When all this has been done, see that springs are strong enough, and not softened by the heat. If they are all right, ship them over their valve stems; if too weak, put in new ones. The cylinder head can now be put back in its place, and bolted on, and we must now see that the exhaust valve opens and closes at just the proper place in the stroke. Whenever we have an oil engine running well, we should always make two marks on the flywheel, so as to show the working of the exhaust valve; then, when we have to put in a new valve, or the old one gets worn, we can always ensure that this most important part of the working of an oil engine is being properly done. If the valve opens too soon, a little must be filed off the end of the stem, until it opens just according to the marks on the flywheel.

When we have done this, we may consider the engine itself as being finished, and the pump and circulating pipes must be overhauled. The packing of the pump should be renewed, and if there was any sign of a leak in any water joints, new washers should be inserted. See that the strainer outside the suction pipe is clear of any obstruction, such as dried mud or weed.

The vaporiser should be opened up, and make sure that the small hole by which the petrol arrives is quite clear, and also that screw valve by which it is regulated is in proper working order. If the latter leaks through its packing, insert new packing well smeared with Sunlight soap, but take great care that no portion of soap gets into the passage for the petrol.

All petrol pipes and tanks should be carefully emptied, and very likely at the bottom of the tanks will be found a small quantity of water. This must be all removed by means of a small sponge, as if it ever finds its way to the vaporiser it will stop the engine. It is best to empty the tanks before we touch the engine, as we can then use the stale petrol thus obtained for cleaning the various parts of the engine.

We must now look at the electric ignition gear. If of the low tension type, see that the sparking points are clean, and that the platinum is not loose in its place in them. See that the rocking arm is quite free to rock, carefully clean every connection in the electric circuit, and having the proper supply of electricity in either our dry cells or accumulators, carefully test throughout the electric circuit, and the spark, and see that they are all right.

If our engine has a high-tension system, see that the platinum contacts in the make and break are quite clean, and that the platinums of the coil trembler are also clean. Test all the circuit, and see that the proper spark is obtained. Having made sure that the spark is of the right sort, see that it is made at exactly the right moment, according to the marks which should always be made on the flywheel, when the engine is new.

A very important thing to see to is that the thrust bearing has not worn so as to put any of the thrust on to the crank of the engine. To test this, unfasten the thrust block, so that the shaft is quite free to move a little fore and aft. Then turn the engine round by hand sharply for about a dozen turns. This will let the crank shaft of the engine take up its natural position. A very accurate mark should now be made on the shaft somewhere, and the thrust bearing must be so tightened up and adjusted that the thrust of the propeller cannot disturb its position, so as to bring any of the thrust on to the engine.

If any bearing used to get hot after a lengthy run, we must find out how the friction arose, and adjust it. If the friction clutch ever used to slip, it should be taken apart and cleaned, and any necessary adjustment, to make the cones bite properly seen to.

New packing should be nserted in the stuffing boxes of the screw shaft, and any unusual play in any part of a reversible-bladed propeller, must be rectified. If any blade is much worn, or badly bent, it will be better to put in a new one, and save further trouble during the busy season.

If all the foregoing have been carried out, we shall be able to get on board the boat, and feel confident that our motor will start off at the first asking, and work on through the season, not only as well as when new, but even better, because, if carefully looked after and adjusted, it will continue to improve with use for a certain time until serious wear and tear begin to assert themselves, and bring about the symptons of that old age which, sooner or later, affects not only the motor but the man who uses it.

CHAPTER XII.

SUMMARY.

ENGINES WITH BOILERS USING LIQUID FUEL — SHIPMAN—L.I.F.U. — AND NAPTHA ENGINES, BOILERS, AND BURNERS—CLARKSON BURNER—ALCO-VAVOUR LAUNCHES.

We now have to pass from internal combustion engines, *i.e.*, those which burn their fuel in the cylinder, to those which burn it in a separate furnace, the heat created thereby being utilised to cause water, or some other fluid, to boil, and thus turn into vapour, the vapour being caused to do work by expanding in a suitable engine. As may be generally expected, these latter engines are not at all so economical of fuel as the former, because there is no doubt that the proper place to burn the fuel with the greatest economy is in the cylinder; but the great advantages obtained by burning it under a boiler may be said to be—(1) Reliability, and (2) Flexibility.

Now the first of these, *i.e.*, reliability, is of great importance in a marine motor, because an unexpected stoppage may be a serious thing, much more so than a stoppage in a road, for instance. If, however, the engine and boiler are of good design and workmanship there is hardly any more risk of a stoppage with liquid fuel than with the ordinary coal-fired steam engine. There is just a little more risk, because the burner using the liquid fuel does introduce a certain amount of non-reliability, but with the latest patterns of burner this is reduced to a minimum, so that we may practically say that the engine is certain to start off at once, when wanted, and to continue to run as long as we want. We can, moreover, run it dead slow, without fearing that it may suddenly stop without warning (as is the case with internal combustion engines). The possession of these two very useful qualifications have rendered the liquid fuel form of engine very popular for certain purposes afloat, such as in launches, etc., and they will probably always hold their own for such purposes.

Another minor advantage is that the engines themselves run quieter than those in which the explosion takes place in the cylinder, and there is no exhaust; but against this it must be said that some of the burners are themselves noisy in use. The difference in consumption in fuel is very marked, for whereas in a well designed internal combustion engine we can get a horse-power hour for just over one pint of petrol, we shall have to burn from 1½ to 2½ times this amount of fuel under the boiler to get the same power, according to the make of boiler and the nature of the liquid within it. With the latest pattern burner, however, we can get good results by burning common paraffin oil, and this, of course, brings down the cost of the fuel used.

The first marine motor using liquid fuel to be seen over here was the American 'Shipman,' introduced many years back, and it embodied many of the principles which have lately been so much talked about and advertised as novelties in motor cars using liquid fuel.

The modern type of burner (Clarkson's) had not been invented, so that the oil had to be *sprayed* by means of steam from the boiler before it could be properly burnt. At first starting, air had to be pumped in by hand until the boiler showed enough pressure to go on spraying the oil, when its action became automatic, because as soon as the pressure in the boiler reached a certain specified amount (which could be chosen and set by hand on a dial) this pressure was used to force a flexible diaphragm against the passage by which the steam went from the boiler to the burner, and cause it only just to keep alight. No more steam, therefore, was thus raised in the boiler, but as soon as the pressure in the boiler fell, either by cooling or by being used in the engine, the diaphragm was withdrawn and allowed the blast to go to the burner, and heat the boiler afresh. Of course the great fault of this system was that the steam was being burnt in the furnace, and steam means fresh water, which is a valuable commodity afloat in salt water. Various attempts were made at subsequent periods to burn the oil by means of compressed air, the pressure being at first supplied by a hand pump, and afterwards by a pump run by the engine. All this, however, gave trouble, and inventors turned their attention towards making a burner which would cause the oil (under pressure) to vaporise itself and burn with the perfect combustion which is so necessary to avoid clogging up the furnace and boiler with soot.

Such a burner was brought over here some few years back and shown in the excellent launches built by the Liquid Fuel Co., of Cowes, I.W., generally called the L.I.F.U. Co., and now established at Poole, and the marvellous speed and perfect running of these engines soon attracted general attention, and a great many launches were built for yachts.

The boiler used is of the water-tube type, *i.e.*, the water is contained in a system of tubes, and the flame plays on the *outside* of them. The water is kept in constant circulation between the lower drum and the upper one, which also contains the collected steam from the tubes. From this upper drum the steam is taken to the engine. It is not proposed here to go into the details of modern engines using steam, as there are so many excellent books dealing with this subject, but it is mentioned as showing what a well-designed engine of this type can do. It weighs 200 lbs., and when supplied with steam at 250 lbs. pressure, it develops 16 h.p.

Coming back to the boiler, at the bottom of the furnace is the burner, a mushroom-shaped apparatus, and the paraffin oil is supplied to it under a constant pressure of about 12 or 15 lbs. per square inch, by means of air being pumped in by hand into the top of the tank containing the oil which is generally in the bow of the boat some way from the boiler, the engine when running keeps up this air pressure. The burner is first of all heated up by means of a torch of asbestos soaked with methylated spirit, and as soon as it is well heated, the paraffin oil is gradually admitted to the winding passages in the head of the burner, and, being vaporised by the heat, escapes in the form of vapour and burns with a large bushy flame, with perfect combustion, so that no soot is deposited on the tubes of the boiler. The heat of this flame not only heats the boiler, and generates steam, but it also just keeps up the requisite amount of heat in the burner itself, to constantly vaporise the incoming stream of oil. It is a very important thing in these burners that the heat should be just right, and not too much or too little. If too great, the paraffin oil tends to get broken up or *cracked* into other elements, and amongst these is tarry matter, which chokes the holes and passages of the burner. Should the heat, on the other hand, get too low, the oil would issue in the liquid form, instead of in the gaseous form, and this should cause a very sooty flame, which would soon coat the tubes over with soot, and thus interfere with the proper steam raising qualities of the boiler. The getting of this exact amount of heat was one of the difficulties with which inventors had to struggle, but in the liquid fuel burner, and in the Clarkson, this has been arrived at in rather different ways. However carefully the heat of the burner is arranged, sometimes a slight deposit of a tarry nature will form somewhere inside it, and being carried along with the vapour, will often choke up any small holes by which the vapour issues; in the L.I.F.U. burner the vapour issues underneath a small safety valve, instead of through a lot of small holes; and, as soon as any deposit interferes with the free issue of vapour the increased pressure lifts up the valve a little, and the deposit is blown out into the furnace. The supply of oil, being regulated by hand, permits of the steam being kept at any steady amount, within limits, when the engine is running, and just before the engine is stopped, the flame is turned down quite low (being just kept alight by a pilot light), all ready to be turned up full blast, just before a fresh start with the engine is required. By attending properly to the burner, the boiler should never blow off through the safety valve, and thus waste fuel and water, however often the enine is started and stopped.

It is really this power of flexibility, more than any other qualification, perhaps, that has rendered the use of liquid fuel so popular for yacht launches. If we are going away for two or three hours steaming at a constant speed, and can give good notice before hand, then a coal-fired boiler is good enough for anyone. But if we want to start in a hurry, and run about, say, calling on yachts or on shore, or dodging about watching a race, or anywhere, in fact, where we require constant bursts of full speed, between periods of rest, then the coal fire does not do. As soon as he knows he is going to stop, the stoker of a coal fire will push his fire back and close the dampers, and even then steam will constantly blow off at the safety valve, with all the consequent annoyance, and, when starting again for the next run, it will take about ten minutes at least to get up the proper fire for anything like full-speed running. All these difficulties, however, disappear with the modern oil burner intelligently used. A launch thus fitted can be run up to a yacht full speed, the fire put out at once, and no one need be left in charge, and this cannot be done with a coal-fired boiler, at all events for ten minutes or a quarter of an hour, during which time the fire would be pushed back, damped with ashes, the dampers closed, and fresh water pumped up into the boiler by the donkey pump.

The invention of a practical burner, using common paraffin oil, with reliability, has given a great impetus to the use of steam for motor cars. In fact, owing to this it is now once more a serious rival to petrol for this purpose, *vide* the Serpollet, Locomobile, Stanley, and other steam cars now so often seen. The illustration Fig 83 shows the Clarkson burner, made by the Clarkson Co. of Great Dover Street, London, which may be said to embody all the most recent improvements in this direction. The oil is kept in a tank under a pressure of air of about 20 lbs. per square inch, and is admitted to the coiled pipe A, A shown on the top of the burner, and, being heated by the flame, becomes vapourised. The vapour is now brought down by a pipe to a jet inside the horizontal main tube of the burner, and close to the end of it, which is closed by a flap E. This flap is kept open in a certain position, and, as the jet of vapour issues under pressure, it induces a draught of air in with it, and the mixture of air and vapour passes up round the curved part of the burner, and issues underneath a conical valve, into what may be termed the combustion chamber of the burner. There it burns with a clear flame free from soot, all ready to heat the tubes of the boiler, and also keep up the proper amount of heat necessary for vapourising the constantly incoming stream of oil from the main tank under pressure.

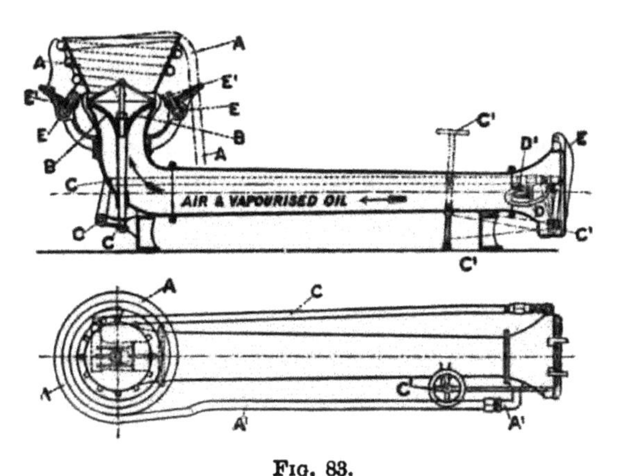

FIG. 83.

CLARKSON BURNER.

One of the greatest features of the Clarkson burner is that of the possibility of regulating the size of the flame, without causing it to become sooty, and this has been a very difficult thing to obtain. It is done, however, entirely by the movement of one lever, which works as follows:—As the lever is moved it either slightly raises or lowers the conical valve in the head of the burner, and by thus either increasing or diminishing the space through which the mixture issues just before it is burnt, the size of the flame is thus accurately controlled. Now, beside the adjustment of the size of the flame, another adjustment is necessary, and that is that of the issuing jet of oil vapour from the vapour jet. This is obtained by making the end of the jet with a *rectangular* hole, and a tapered rectangular pin is caused to move slightly in or out of this jet hole as the lever is moved, the whole apparatus being so accurately made and adjusted that the amount of oil vapour issuing from the jet is always exactly what is necessary for the size of the flame required. It is worthy of note that if the jet had a *conical* hole and a *conical* pin, the two adjustments would not be always right, but with the rectangular section for the hole and pin they are so.

In order to start from all cold a certain measure of methylated spirits of wine is poured in through a special pipe, and the spirit is lit in the head of the burner. In a few minutes the head of the burner and the coiled pipe become hot enough to allow of the oil being slowly admitted from the pressure tank, and the oil becomes vapourised, and this vapour passes down through its pipe to the jet in the main tube of the burner, and passing up, together with the proper amount of air, is lit in the combustion chamber by the still burning spirits of wine.

The flame in the burner is sheltered from the effects of wind by outside perforated metal guards, as well as by a special coil of wire, and this latter also acts as a silencer, and renders the burners quiet in use.

The various paraffin oils when used as liquid fuel have one great indisputable advantage over coal, *i.e.*, that they have, weight for weight, and bulk for bulk, far greater heating, and consequently, steam-raising power. Careful experiments have now been made in this country extending over many years with liquil fuel, in steamers, locomotives, and stationary boilers, and as a result of them, it may be stated that with the cheaper and heavier forms of paraffin oils, of a specific gravity of ·92, the relative heating value of 1 lb. of oil as compared with 1 lb. of the best anthracite coal, is as 11,200 to 8,458; but taking into account all the complicated chemical changes which take place in a furnace, the practical outcome is that there is a saving when using oil of about 40·5 per cent. in weight, and about 36·5 per cent. in bulk over coal. These figures apply, as stated above, to the heavier refuse oils which are left after the usual paraffin oil of commerce has been distilled off. At present there is no regular market for these heavy oils in England, but should the demand ever spring up, no doubt they would be brought over here, and they could then be sold at a fraction of the price of the usual paraffin oil.

As these heavier oils, however, are of a more or less sticky nature, it is very doubtful if they could be burned without either a steam or air jet. But, as the heating power of ordinary paraffin oil, compared with that of the heavy oil, is as 10,192 to 11,200, we can see that the use of the former in the modern type of self-acting burner (Clarkson's) gives us certain great advantages in the weight and bulk of fuel we have to carry in a boat, as compared with coal.

Many of our readers may remember a peculiar type of launch from America that used to be seen in the window of Messrs. Rowland Ward, of Piccadilly, and they were also constantly seen afloat (as indeed they still are). These are called *naphtha launches*. They were called naphtha launches because they use naphtha, or what we now call petrol, *inside* the boiler, and as it was heated it turned into vapour under pressure, just like steam, and this compressed vapour was used to drive a small engine of the simple steam engine type. Having done its work in the cylinder, it was turned into a condenser tube running along the length of the keel under water, and becoming thus condensed again into liquid naphtha, it was pumped back again by the engine into the boiler for use again. A certain portion of the vapour was led into a burner under the tubes of the boiler, where it was burnt, thus keeping up the necessary boiler heat. As this flame could be easily regulated by hand, any pressure of steam could be kept up in the boiler, according to the requirements of the moment.

At first starting, air had to be pumped by hand for some minutes, and the retort over the burner heated by a spirit lamp; when sufficiently heated the burner became automatic in action, and the pressure in the boiler rapidly rose to that necessary for working. In hot weather the launch could be started from all cold in about three or four minutes.

The great advantage claimed for this system was the saving in weight over the usual steam plant, whilst preserving the reliability and flexibility of the latter. Thus the weight of the whole equipment for 2 h.p. is given as being only 200 lbs., and that for an 8 h.p. plant at 600 lbs.

Now the saving in weight depends on the fact that a given quantity of heat will evaporate a greater quantity of petrol than water, as will easily be understood, but the pressure of the petrol vapour will not be equal to that of the steam; but petrol vapour has this advantage over steam, that it can be expanded down in the working cylinder to a temperature of 130° F. before it is condensed again to the liquid form, whereas steam, we know, must not be allowed to get even near the temperature of 212° F. or it will condense in the cylinder. The practical outcome of these pros and cons is that a given quantity of heat will produce, say, 9 h.p. hours with petrol in the boiler, where it would only produce 5 h.p. hours with water in the boiler.

Messrs. Yarrow of Poplar, made some very exhaustive and instructive experiments with *spirit boilers*, as they were called, and the practical results of these experiments agreed almost exactly with the proportions of efficiency as calculated by theory, *i.e.*, 9 to 5.

The reason why these launches never became very popular over here was probably because people did not particulary like the idea of having so many gallons of boiling naphtha on board under a considerable pressure; and, moreover, the amount of this material burnt in the furnace was con-

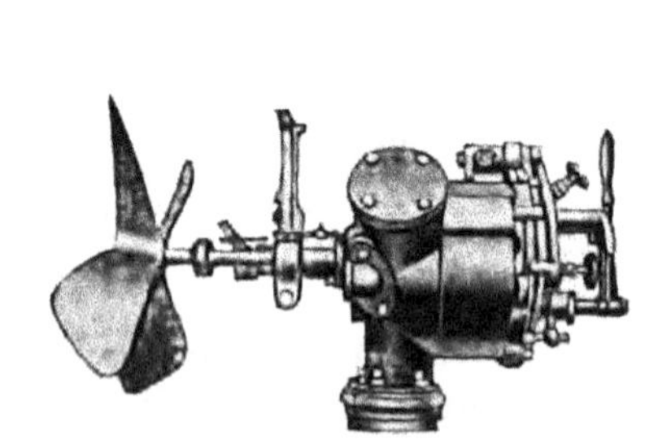

FIG. 84.
ALCO VAPOUR ENGINE.

FIG. 85.
ALCO VAPOUR ENGINE AND BOILER.

siderable in a day's running, and the boiler was constantly being replenished from a tank in the bow to make up for this loss. Messrs. Yarrow tried different ways of burning common paraffin oil, by air jets, etc., but the modern type of Clarkson burner had not been invented.

What is the most modern application of this system is that used in the Alco. Vapour Launches, and they possess certain clear advantages over the oil spirit launches described above, and they are much used in America, where they were invented; (they used methylated spirits of wine in the boiler and paraffin oil as fuel).

Fig. 84 shows the engine connected to the propeller, and Figs. 85 and 86 show the boiler and machinery complete, which, as will be seen, is of a very compact and neat appearance, and consists of a small three-cylinder engine placed below a tubular boiler. The cylinders are placed 120° apart, so that there is no dead point, and the engine can be reversed at once when running full speed. The chief advantages claimed for the use of alcohol inside the boiler, as compared with naphtha or petrol, are (1) that alcohol produces more vapour from a given quantity of heat, and consequently does

Fig. 86.
Alco Vapour Machinery.

more work, and (2) the peculiarity that alcohol has of mixing with water, instead of floating on its surface like petrol. Thus, supposing a leak took place, and the alcohol were to run down into the boat, it will mix with any bilge water there, instead of floating on its surface; and, moreover, if a bucket of water were thrown over a pool of burning alcohol they would mix together and the flame would be extinguished. In case the supply of alcohol happened to run out, it is possible to use water in the boiler, and thus be able to continue a passage. Of course this would only be done on an emergency, and only a slow speed could be thus obtained.

As will be easily understood, the equipment of machinery with this system is one of the lightest for power; and as the only alcohol actually used is that lost through leaky joints, pumps, etc., this should not amount to much in a season's running with a careful engineer. The following are the sizes of standard equipments, *i.e*, 1, 2, 3, 5, 7 and 12 h.p. The sole agent in England is Mr. George Wilson, of Sherwood Street, Piccadilly, London.

It is, perhaps, rather curious that yachtsmen generally do not seem to have taken to heart the lessons which the large sailing ships of the mercantile marine and the great fishing fleet of the North Sea have been teaching us for years, *i.e.*, that it is of great advantage to have some form of power available for special hard work, and thus do away with some of the crew. Many a large sailing ship is now afloat, which more or less entirely depends on a steam boiler for the periodical hard work entailed by the use of sails; and there is now hardly a trawlerman in the North Sea without his steam capstan and donkey boiler, and even the drift net boats are taking to them. It can hardly be doubted that before very long yachts will be supplied with some sort of power, and will consequently be able to do with smaller crews, and this could not fail to be of the greatest benefit to all concerned with the yachting interest, because any great saving in the expense of yachting would have an enormous influence on the number of yacht's built and used. Crews' wages are the most expensive item in the season's yachting, and it is natural, therefore, to look for some saving under this head. Formerly, one always had to be able to turn out a strong boat's crew of four or five oars, whilst leaving enough hands on board to handle the yacht properly, but now that small power launchns are comming in two men are quite sufficient for all landing work.

The form of power suitable for assisting in the work of a yacht would probably be a neat and compact form of capstan worked by one of the types using liquid fuel described in this chapter: and it would be an interesting thing to see such a capstan brought out on the market.

CHAPTER XIII.

SUMMARY.

ELECTRIC MOTORS—GENERAL PRINCIPLES AND ADVANTAGES—DISADVANTAGES—THAMES VALLEY LAUNCH CO.—WESTMACOTT AND STEWART—DETACHABLE ELECTRIC PROPELLER—GENERAL REMARKS—FINAL.

The only other type of marine motor now remaining to be described in this book is the electric motor, and all those who use the River Thames are well acquanted with the appearance of the splendid fleet of electric launches built by the Thames Valley Launch Co., of Weybridge; and no one can fail to admire the quiet running, perfect control, and freedom from noise and smell which characterise these motors so especially. Their patent system of using only one lever for steering, and also controlling, the motor, renders the launches suitable for the use of ladies, and even children. It is, however, chiefly on inland waters that these launches are so much in evidence, and they are not often seen on the sea, and the reasons for this will be given later on.

Commencing with the actual motor, it may be truly said that the electric motor, as now made, is the most efficient motor that we have, as regards the amount of energy absorbed and lost in its working. Modern electric motors will give out 80, and even 90 per cent. of the energy of the electric current supplied to them, and this means an efficiency many times greater than is possessed by any of the other motors described in this book.

In order to follow out the working system of the electric motor, we must refer back to Chapter VIII., where some of the rudimentary principles of electricity were described and explained. It will be remembered that there it was stated that whenever an electric current was sent through a coil of insulated wire which surrounded a mass of soft iron, or bundle of soft iron wires, the soft iron became magnetised, during the passage of the current; but, as soon as the current stopped, the iron lost all its magnetism. Now anyone who has handled any of the usual horseshoe magnets, as usually sold, must have noticed that *dissimilar* poles *attracted* each other, whilst *like* poles *repelled* each other; that is, the pole marked N will attract any pole marked S, but will repel any pole similarly marked N. Referring now for the moment to Fig. 87, which shows a magneto sparker for oil engines, we see that it consists of a revolving armature turning between the poles of the fixed magnets. This armature consists of a core of soft iron surrounded by a coil of insulated wire, so that if we were to connect up the ends of this wire with a battery, and thus send a current through it, we should magnetise the soft iron armature, and one of its sides would develop strong N magnetism, whilst the other side

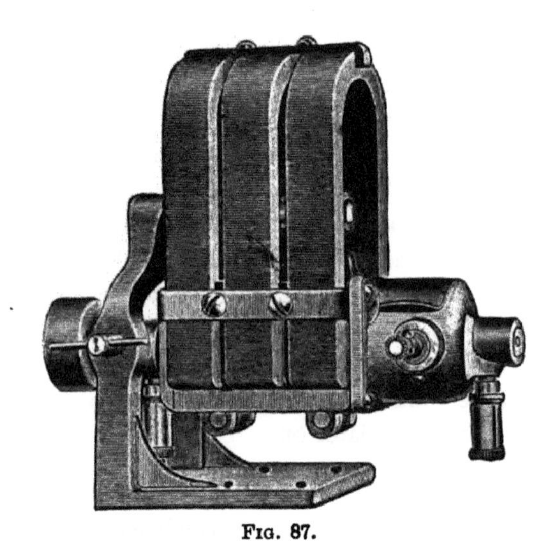

Fig. 87.

Magneto Sparker.

would develop equally strong S magnetism. Now, supposing that the N pole of the armature were opposite the N poles of the fixed magnets, and consequently the S pole of the armature would be opposite their S poles, both poles would be repelled, and the armature would fly round, and, getting an additional pull from the attraction of the opposite poles, would come to rest with its N and S poles just opposite the S and N poles respectively of the fixed magnets. Supposing, however, that just as the armature reached this position, the electric current were reversed so as to reverse the poles of the armature, it would then fly round another half circle, and if we arranged the commutator so as to constantly reverse the current at exactly the right moments, we would find that the armature would keep spinning round, and would, moreover, give out a certain amount of power or energy, which latter would depend on the strength of the current given out by the battery. We should now, in fact, have turned, our magneto sparker into an electric motor. There are a great many systems of making and wiring electric motors. Some of them possess many more poles than two; and the fixed magnets are replaced by other electro magnets, which give more powerful results; but they all work on the general principle we have just described. For a given power, they are a little heavier than the best petrol motors, but we have mentioned their extraordinary efficiency; and the perfect control which we have over their movements, the power of reversing, the smooth running, and freedom from noise and smell have rendered them the ideal type of motor for inland waters.

When we come, however, to the means of supplying the motor with the necessary current of electricity, we put our finger on the weak point of the electric system as compared with others. We are obliged to carry about in our boat a bulky and very heavy arrangement of accumulators. In Chapter VIII. we gave a description and drawing of the accumulators, as used for electric ignition, and the larger ones used for electric motors work on just the same principle. Different makers of electric motors use different amounts of current, but 100 volts is a very usual pressure, and, as each cell of an accumulator gives a pressure of 2 volts, they are usually connected up in sets of 50 cells, in series. Some boats carry a double set of such cells, so that, as one set becomes exhausted, the other fresh one can be switched on to do the work. The size of the plates regulates the quantity of current given out, so that the larger they are the more hours work we can get out of them. It is very unfortunate that the best accumulators we can make have their plates made of lead, as this, being a heavy metal, makes the accumulators very heavy for a given power. It is customary to speak of the storage energy of an accumulator in *Watt hours,* and we must just describe the meaning of the terms a little before using them generally. A Watt is one volt multiplied by one ampère. Thus, an accumulator giving a current of 20 ampères at a pressure of 2 volts is giving out 40 Watts, and if it could continue to do this for five hours before becoming exhausted, its storage capacity would be said to be 200 Watt hours.

Seven hundred and forty-six Watts are equal to 1 h.p., so that if we have an accumulator which will give a current of 7·46 ampères at a pressure of 100 volts, it will give out energy equal to 1 h.p., and if it could continue doing this for, say, one hour, before becoming exhausted, its storage capacity would be said to be 1 h.p. hour, or 746 Watt hours.

Coming now to the actual weight of accumulators on the market, we find that the average accumulator weighs from 124 lbs. up to 186 lbs. per horse-power hour, so that it will now be seen what an enormous weight we have to carry about in a boat that is to go any distance without recharging. Another great nuisance is that accumulators cannot be recharged all at once. This must be done slowly, and takes therefore many hours ; generally about six hours suffice. A third drawback exists in the fact that there is always more or less leakage going on, even if the boat be not used. These drawbacks will explain why the electric motor is not so much used at sea as on inland waters such as the Thames, where there are plenty of charging

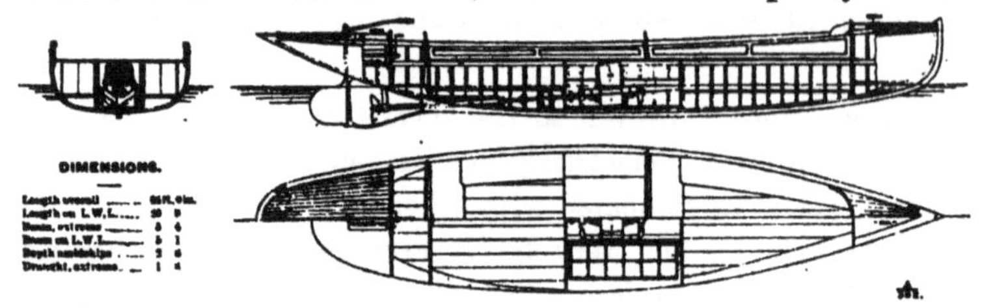

PLAN OF 26 FT LAUNCH.

FIG. 88.—ELECTRIC SEA-GOING LAUNCH. (WESTMACOTT & STEWART).

stations always handy, and the launch can be tied up for the night at any one of them, and is all ready charged next morning for another day's run.

Electric launches, however, are very useful for carrying in large steam yachts where a dynamo is fitted for recharging them, and this latter operation can be done either in the davits or whilst lying astern of the yacht. Such a launch is always ready for the owner or his family at any instant day or night. In some yachts the accumulators of the launch are used to light the yacht with electric light, and thus the noise of the dynamo is done away with at night.

Fig. 88 shows a 26-ft. seagoing launch as constructed by Messrs. Westmacott & Stewart, of St. Helen's, I.W., who have supplied so many of this type to yachts, and also the large seagoing steamers. The accumulators are carried in special watertight boxes on either side of the machinery, and the launch can go for five hours at full speed, or 14 hours at half speed, without recharging.

Fig. 89 shows an ingenious form of electric propelling rudder, which can be instantly shipped or unshipped as required. The motor (which has a watertight cover, not shown) is supplied with the electric current from accumulators carried in the boat, and it drives the submerged screw by means of a belt which runs down in a watertight case so that it is never wet. The weight of the rudder complete, as shown, is about 40 lbs. for

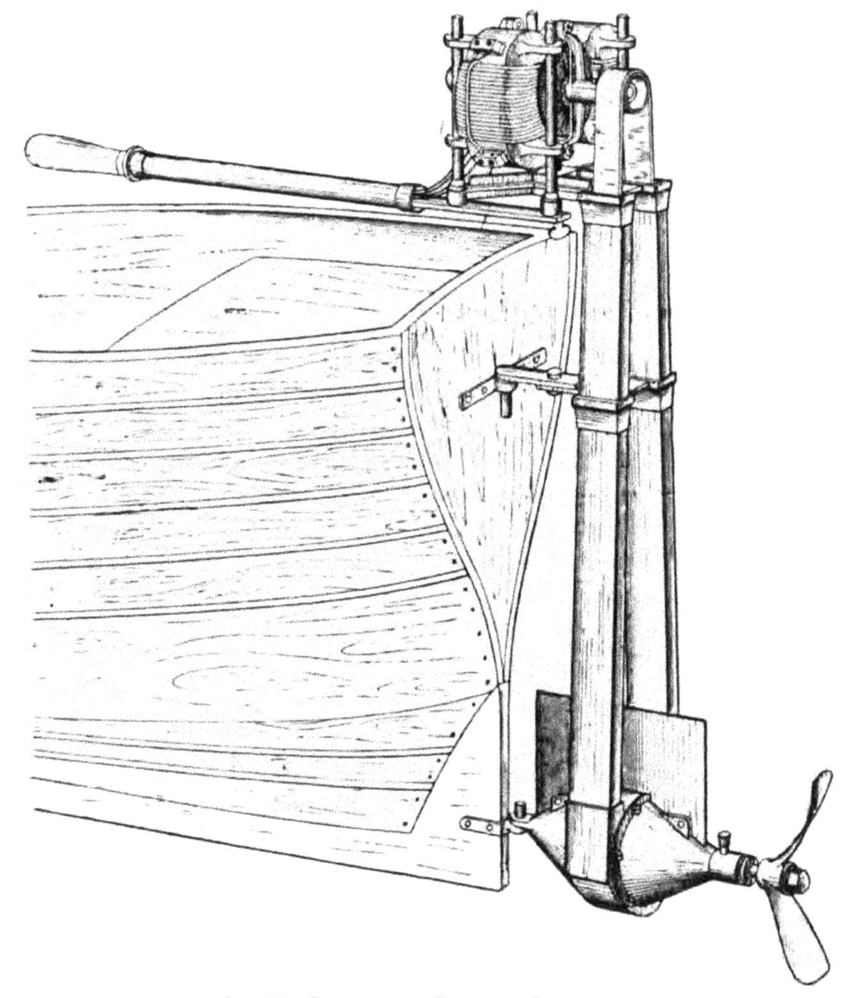

FIG. 89.—DETACHABLE ELECTRIC PROPELLER.

$\frac{3}{4}$ h.p. It is sold by the United Motor Industries, of 64, Holborn Viaduct. Another form of similar electric rudder is made by Mr. E. A. Maclachlan, of Chiswick.

Accumulators are not nice to carry under the decks of a yacht, unless the compartment in which they are stowed is quite cut off from the rest of the yacht, and is also provided with proper ventilation to the outer air. When charging acid fumes and hydrogen gas are given off, and whilst the former are naturally bad for breathing and decorations, the gases may form an explosive mixture with the air.

Inventors naturally have been endeavouring for years to get a good accumulator of lighter weight, and the last invention of Edison in this direction appears to offer a chance of this being done. In this invention, instead of the heavy metal lead being used, the plates are made of nickel, carbon and spongy iron, and the inventor states that he can get a horse-power hour with only 53 lbs. weight. It must be noted, however, that the average pressure of the current is only just over 1 volt per cell, so that about double the number would have to be carried for a given voltage, and, although a saving in weight may be obtained, probably the bulk would be greater. Of course, in a boat the bulk is of great importance.

A German scientist has recently made some very interesting discoveries by heating the accumulators whilst drawing on them for current, and he states that he has succeeded in getting nearly double the power out of a given weight by this means. The ideal which all electricians have been seeking for years is to produce the electric current direct from heat, and if this could be done economically, we should have a perfect form of motor, not only for marine purposes, but also for land transport. There are several ways known of thus producing electricity direct from heat, but they are too wasteful to be employed practically.

With these few remarks on electric motors this book must now conclude, as in it has been given a short descriptive account of all the most divergent types of marine motors now on the market. Every endeavour has been made to render the descriptions as impartial as possible, though perhaps those interested in some special type may think that enough space was not devoted to it.

If a study of this work has been the means of persuading anyone either to buy a motor launch, or to put a motor into his own boat, it will not have been written in vain. No yachting man who has not been through it can imagine the feeling of delight and pride when he takes the tiller of his favourite sailing boat, and feels for the first time that he can not only glide about in a calm, but that in a sailing breeze, when the circumstances of navigation demand it, he can actually enter and handle his boat within the eight forbidden points of the compass, which, as a sailing man, he has all his life regarded as a *mare clausum*.

THE END.

Lightning Source UK Ltd.
Milton Keynes UK
UKHW052350191118
332601UK00007B/460/P